EGRESS

FOREWORD BY

Stephen Hawking
and Lucy Hawking

EDITED BY

Robert Jacobs,
Michael Cabbage,
Constance Moore,
and Bertram Ulrich

WITH IMAGES SELECTED
BY THE APOLLO ASTRONAUTS:

Buzz Aldrin
William A. Anders
Neil A. Armstrong
Alan L. Bean
Frank Borman
Eugene A. Cernan
Michael Collins
R. Walter Cunningham
Charles M. Duke Jr.
Richard F. Gordon
Fred W. Haise Jr.
James A. Lovell Jr.
Thomas K. "Ken" Mattingly II
James A. McDivitt
Edgar D. Mitchell
Harrison H. "Jack" Schmitt
Russell L. Schweickart
David R. Scott
Thomas P. Stafford
Alfred M. Worden
John W. Young

APOLLO

THROUGH THE EYES OF THE ASTRONAUTS

ABRAMS, NEW YORK

Contents

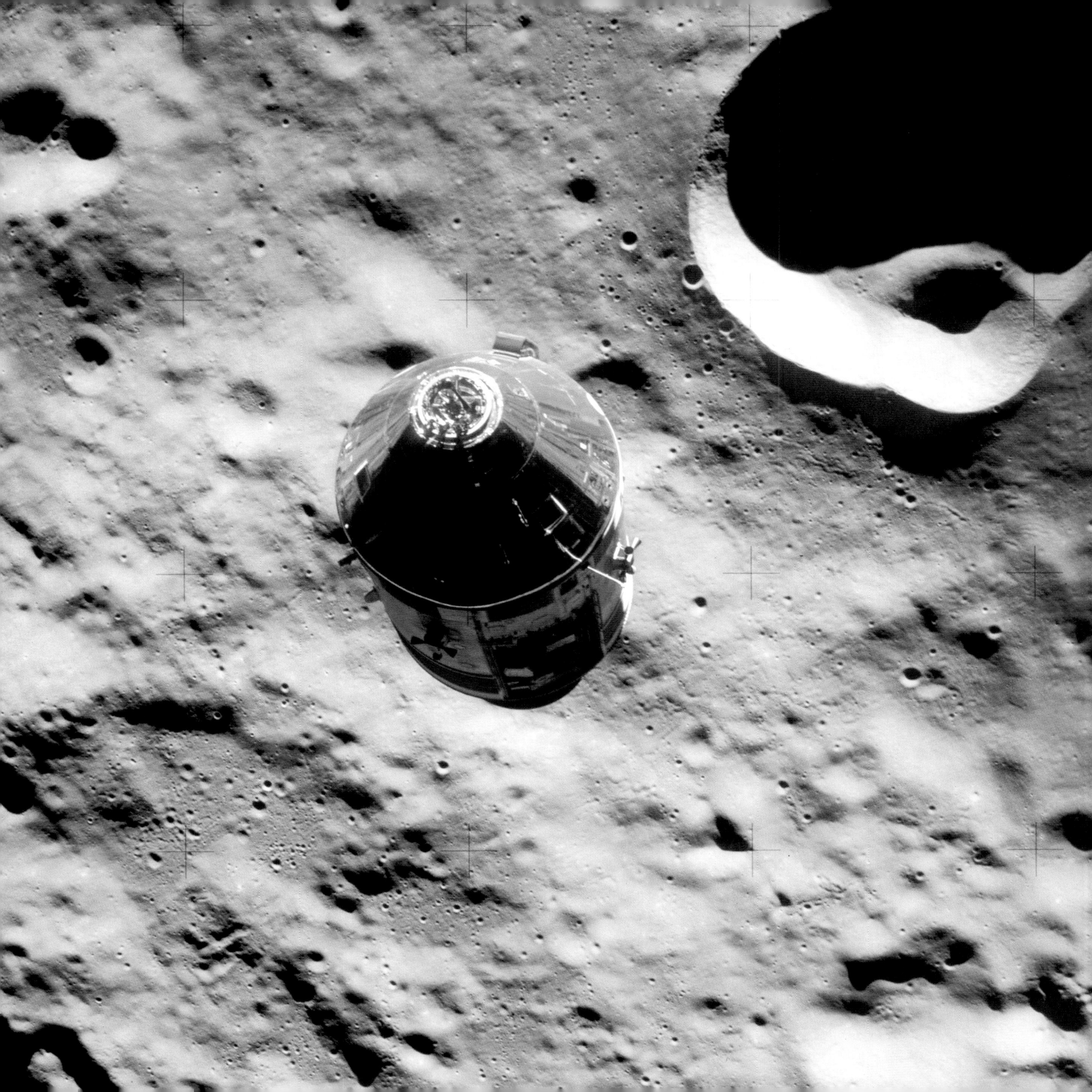

PAGE 1
During the Apollo 11 prelaunch countdown on July 16, 1969, Neil A. Armstrong and Michael Collins arrive at the summit of Pad A on their way to the spacecraft that would carry them to the moon.

PAGE 2
The three Apollo 14 astronauts enter the white room, the chamber at the top of the launch tower that gives access to the command module. Commander Alan Shepard can be identified by the red bands on his spacesuit. Astronaut Thomas P. Stafford, Chief of the Manned Spacecraft Center Astronaut Office, faces the camera. January 31, 1971.

PAGE 3
In the white room, Apollo 10 astronauts are preparing to enter their spacecraft on May 18, 1969. Lunar module pilot Eugene A. Cernan is standing in the right foreground; Thomas Stafford, commander, is at left; in the background, a technician kneels in front of the open door of the command module *Charlie Brown*.

PAGE 4
Lightning flashes in the sky behind the Saturn V rocket that will propel Apollo 15 to the moon, July 25, 1971.

PAGE 5
The fisheye lens of a camera mounted on the mobile launch tower captures the liftoff of Apollo 11, July 16, 1969.

PAGES 6–7
On December 7, 1972, the 363-foot-tall Apollo 17 space vehicle lifts off from Pad A at the Kennedy Space Center in Florida, marking the first and last night launch of the Saturn V.

PREVIOUS SPREAD (LEFT)
A distant command service module floats above the lunar surface during the Apollo 17 mission. Photograph taken from the lunar module *Challenger*, December 11, 1972.

PREVIOUS SPREAD (RIGHT)
The command service module as seen from the lunar module *Orion*, which was beginning its descent to a landing in the lunar highlands, Apollo 16, April 20, 1972.

OPPOSITE
The Apollo 12 lunar module *Intrepid* is set in a lunar landing configuration. Photograph by Richard F. Gordon, November 19, 1969.

Foreword

Stephen Hawking and Lucy Hawking

We thought space was worth a big effort in the 1960s. In 1961, President John F. Kennedy committed the United States to landing a man on the moon by the end of the decade. This was achieved just in time by the Apollo 11 mission, in 1969. The historic images captured by the Apollo astronauts, some of which are featured in this book, helped fuel our fascination with science and technology.

Why did we go into space? What was the justification for spending all that effort and money on getting a few lumps of moon rock? Weren't there better causes here on Earth?

In a way, the situation was like that in Europe before 1492. People might well have argued that it was a waste of money to send Christopher Columbus on a wild goose chase. Yet, the discovery of the New World made a profound difference to the old. Sending humans to the moon may yet prove to have had an even greater effect. It changed the future of the human race in ways that we don't yet understand and may have determined whether we have any future at all. It hasn't solved any of our immediate problems on planet Earth, but it has given us new perspectives on them and caused us to look both outward and inward.

After the last moon landing in 1972, with no future plans for further manned space flight, public interest in space and willingness to fund further exploration declined. This went along with a general loss of confidence in science in the West because, although it had brought great benefits, it had not solved the social problems that increasingly occupied public attention.

We need to renew our commitment to human spaceflight. Robotic missions are much cheaper and gather important scientific data, but they don't spread the human race into space, which should be our long-term strategy. If one is considering the future of humanity, we have to visit other worlds ourselves. Exploring other worlds won't be cheap, but it will take only a tiny proportion of this world's resources. NASA's budget has remained roughly constant in real terms since the time of the Apollo landings, but it has decreased from about 4 percent of the U.S. federal budget in 1970 to 0.6 percent now. Even if we were to increase the international budget twenty times to make a serious effort to go into space, it would only be a small fraction of world resources. There are those who argue that it would be better to spend our money solving the problems of this planet, like climate change and pollution, rather than wasting it on a possibly fruitless search for new planetary homes.

We do not deny the importance of fighting for a better Earth, but we can do that and still spare a few percent for space. Isn't our future worth it?

A new space program would also increase the public standing of science. The low esteem in which science and scientists are held is having serious consequences. We live in a society that is increasingly governed by science and technology, yet fewer and fewer young people want to go into science. The children of today are the adults of tomorrow, and they need to have a basic understanding of science and technology if they are going to be prepared to make the hard decisions on issues like fuel sources and food production that affect us all. We've talked to kids around the world and discovered an enormous appetite and enthusiasm for knowledge about space exploration and science. A significant percentage of students studying sciences at a higher level report that their early interest was sparked by exactly these kinds of topics. Space has the power to ignite children's imaginations and engage their curiosity. There seems absolutely no doubt that we have never needed to do this more urgently.

Many of today's space explorers watched the Apollo moon landings in their pajamas, gazed with childhood fascination at the photographs taken, and decided they were going to grow up to extend humanity's reach to other worlds. Others looked at NASA's photographs of Earth from space and saw what a beautiful planet we live on, but also how vulnerable it is, and made a stronger commitment to look after it. These images influenced their life choices.

What's next? Can we exist for a long time away from Earth? The International Space Station proves that it is possible for human beings to survive for many months away from planet Earth. Any long-term base for human beings, however, will probably need to be on a planet or moon, where subsurface enclosures can provide explorers with thermal insulation and protection from meteors and cosmic rays, and the raw materials that are necessary to create self-sustaining extraterrestrial communities can be found.

What are the possible sites of a human colony in the solar system? The most obvious is the moon. It is close by and relatively easy to reach. As illustrated in this publication, we have already landed on it and driven across it in a buggy. On the other hand, the moon is small and without an atmosphere or a magnetic field to deflect solar radiation. There is no liquid water, but there may be ice in the craters at the north and south poles. A colony on the moon could use this as a source of oxygen, with power provided by nuclear energy or solar panels. The moon could become a base for travel to the rest of the solar system.

Mars is the obvious next target. It is half as far from Earth as Earth is from the sun and so receives half the warmth. Its magnetic field decayed four billion years ago, stripping the Red Planet of most of its atmosphere, making it impossible for liquid surface water to exist. However, the atmospheric pressure must have been higher in the past because we see what appear to be runoff channels and dry lakebeds. This suggests that Mars had a warm, wet period during which life might have appeared. There is no sign of life on Mars now, but if we found evidence that life had once existed there, we would know that the probability of life developing on a suitable planet was fairly high throughout the universe.

What about beyond the solar system? Our observations indicate that a significant fraction of stars have planets around them. So far, we can detect only giant planets like Jupiter and Saturn, but it is reasonable to assume that they are accompanied by smaller Earthlike planets. Some of these will orbit in a "Goldilocks Zone" where conditions permit liquid water to exist on their surface.

There are approximately a thousand stars within thirty light years of Earth. If 1 percent of each had Earth-size planets in a Goldilocks Zone, we would have ten candidate new worlds. We cannot reach them with current technology but we should make interstellar exploration a long-term aim. By long term, we mean over the next two hundred to five hundred years, and by exploration, we mean with humans.

The human race has existed as a separate species for about two million years. Civilization began about ten thousand years ago, and the rate of scientific and technological development has been steadily increasing. If the human race is to continue for another million years, we will have to boldly go where no one has gone before. The images featured in the following pages are merely a preview of what's to come.

OPPOSITE
The Earth moves directly between the sun and the Apollo 12 spacecraft in this solar eclipse photo taken during the crew's journey home.

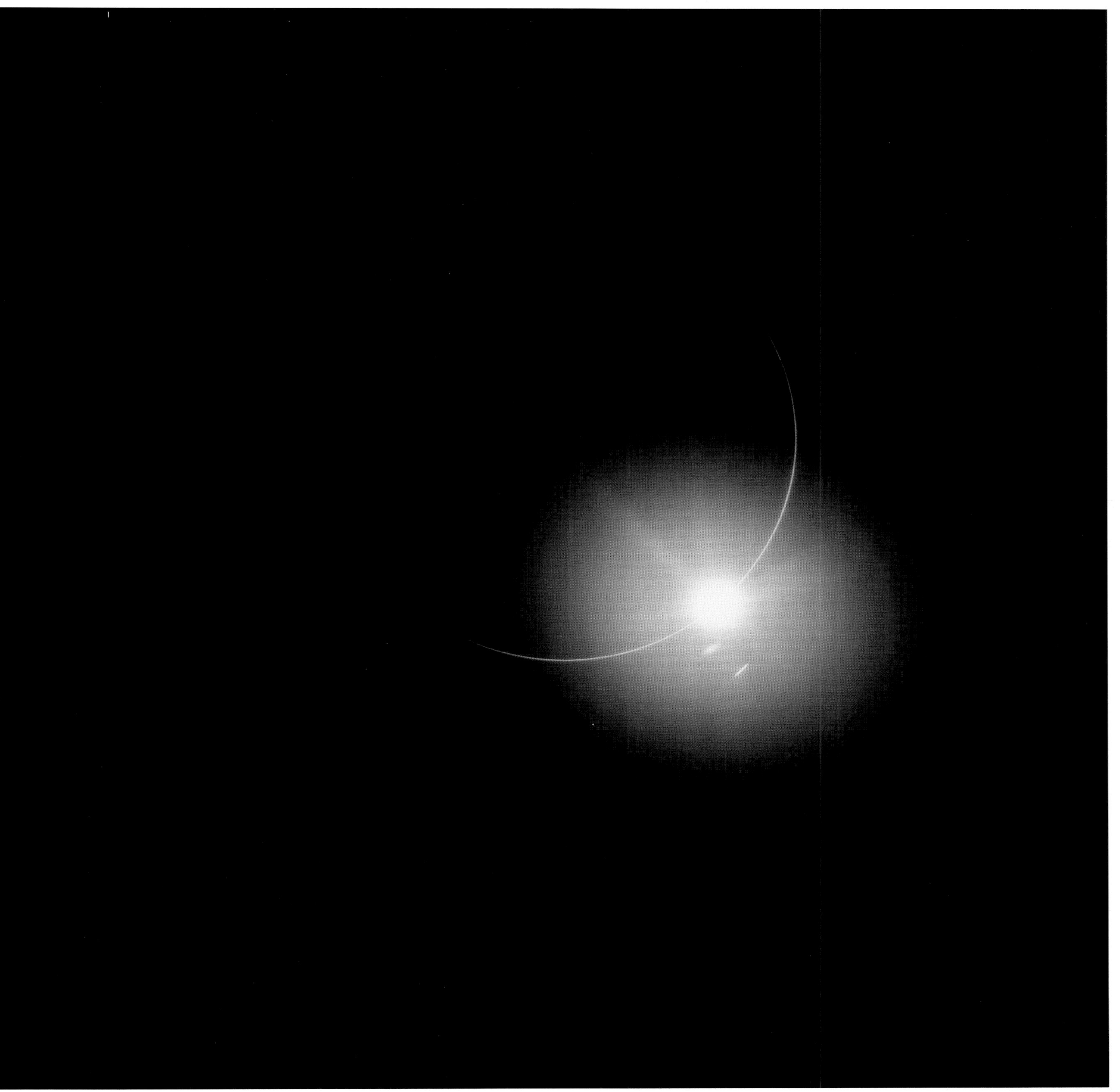

MAIN MOM
TALK

Introduction

Since the invention of the modern camera in the 1830s, no one has captured more on film about humankind's first forays into the cosmos than twenty-nine amateur American photographers whose entire portfolio spans the five short years from 1968 to 1972. They were test pilots, engineers, and one geologist. Primitive by today's standards, their cameras produced some of the most celebrated images ever taken. Their canvas? A desolate landscape on an alien world.

Three of those photographers, astronauts Neil Armstrong, Buzz Aldrin, and Michael Collins, first saw the pictures they captured during their landmark Apollo 11 moon landing while quarantined inside part of Building 37 at NASA's Johnson Space Center in Houston. It was several days after splashdown in July 1969. Johnson's space photography manager, Dick Underwood, used a slide projector to show the images through a window to the trio inside. There initially had been plans to quarantine the film, just like the crew, to make sure no potentially lethal germs had hitched a ride back from the moon. But interest in quickly getting the photos out to an anxiously waiting world won out over any lingering concerns.

Unlike the digital photography of the twenty-first century, where amateurs and professionals alike can review the fruits of their labor seconds after pressing the shutter button, the Apollo 11 astronauts had no way to know in advance whether the pictures were any good. However, Collins, who orbited the moon while Armstrong and Aldrin descended to the lunar surface, remembered his crewmates had expressed cautious optimism on the voyage home that the images were something special.

"Neil and Buzz both were saying, 'We got some great pictures—I hope, I hope.'"

As the three men's photos were projected inside the quarantine trailer a week later, they weren't prepared for what they saw.

"The pictures were of such clarity and startling starkness," a still-impressed Collins recalled nearly four decades later. "I thought 'Not only did we do this, but we brought back amazing documentation.' These weren't grainy black-and-white photos. You could see every rivet, every fold in the flag."

Today, some of the mission's 1,407 70-mm images rank alongside the Iwo Jima flag-raising as among the most iconic American photographs of the twentieth century. There is Aldrin saluting the flag; the footprint in the lunar dust; Armstrong's photo of Aldrin that captured the lunar lander, the flag, and himself in his crewmate's visor. Four decades later, Aldrin still points to Apollo 11's body of photographic work as a milestone.

Aldrin said recently, "Nothing can replace the pictures that were taken on the first visit to the lunar surface. Those are always going to stand out."

The photography from the ten other manned Apollo missions are every bit as remarkable. As a whole, the images compare to those of celebrated American photographers Mathew Brady, Margaret Bourke-White, Dorothea Lange, and Ansel Adams in terms of social and historical impact. When first released during the turbulent times of the late 1960s and early 1970s, the pictures instantly transcended cultural, political, and international boundaries—and continue to do so today. One of the images, the famous Earthrise picture snapped by Apollo 8 astronaut Bill Anders on Christmas Eve 1968, was called "the most influential environmental photograph ever taken" by renowned wilderness photographer Galen Rowell.

The goal of this book was to collect the favorite Apollo images of the men who took them and, for the first time, present many of the stories behind the images. In some cases, the selections were surprising. Many of the images might not have been taken had the astronauts not fought behind the scenes to keep photography from being curtailed for other mission priorities.

"This battle went on and on, but we always had the support of the astronauts," Underwood said. "I remember on Apollo 16 when they wanted to cut a number of [film] magazines off the flight in a meeting. These guys were saying, 'Well, photography is no good. What difference does it make? That's not important.'"

"[Apollo 16 Commander] John Young, being the type of guy he is, I remember said to them, 'You know, I know a lot of people that go on vacation, and they go to some great places. They take a lot of photographs to bring home, but I don't know a single one of them that went on a vacation and came home and was kicking themselves because they took too many photographs.' We got the extra magazines."

The photographic tools used by Young and the other Apollo astronauts were archaic compared to today's digital, high-tech tools. Moon walkers on the lunar surface relied mainly on specially modified 70-mm cameras manufactured by the Swedish company Victor Hasselblad AB. The Hasselblad cameras were semiautomatic, battery powered, and produced quality still photographs with minimal adjusting of the settings. Much of the time, the cameras were mounted on chest-high brackets attached to the front of the moon walkers' spacesuits. Astronauts took pictures by squeezing a trigger on the camera's handle.
The missions carried a stockpile of film magazines that could easily be installed inside the Hasselblads. Some magazines contained color film. Others used black-and-white. In addition to the Hasselblads, a 16-mm data-recording camera took motion pictures.Training was relatively simple and informal. Astronauts were given the cameras with instructions on the proper use of settings. Most practiced by taking family and vacation photographs, which were developed and critiqued by Underwood and others. Some crews, such as Apollo 11's, took pictures of the ground from their jets as they flew back and forth between Houston and Cape Canaveral for training exercises. Moon walkers also practiced at several locations, including a mock lunar landscape at the Kennedy Space Center (KSC) in Florida. Apollo 16 moon walker Charlie Duke recalled recently how the exercises were a point of pride for astronauts who already had myriad other critical responsibilities on their minds.

"We worked hard at remembering the different camera settings that were given to us preflight," Duke said. "We trained hard at it. We practiced it in the suit generally down in Florida. We got pretty good at pointing the camera."

Apollo 15 moon walker Dave Scott said the hours of practice on Earth made photography on the lunar surface not much harder than snapping pictures on vacation. His mission returned more than 2,600 Hasselblad images to Earth.

"When we got on the moon, even though we had to operate the camera manually and didn't have any range finder or viewfinder or light meter or anything like that, because we had the opportunity to use the camera and be critiqued on it, it was pretty easy," Scott remembered. "Of course, you had to stand still. You had to make sure your body wasn't moving when you took the picture, and the sun angle was right and the lighting was right and the depth of view was right and all that. But having had the opportunity to use the camera, why, it was just basically mechanical."

The training paid off more than anyone could have expected—the eleven manned Apollo missions yielded 18,667 Hasselblad photos. However, luck and teamwork also played a role in many of the pictures. The famous Earthrise image is an example. For a time, there was debate about which of Apollo 8's astronauts took the picture of a sparkling blue Earth rising above the desolate lunar landscape. All three crew members had begun furiously snapping photos when Jim Lovell spotted the unprecedented opportunity. Later, it was determined that Anders got the shot.

"We have always had arguments over the years of who took it," Lovell chuckled four decades later, "but in reality, it was Anders who took the picture because he had a telephoto lens on his camera. If you look at the picture of Earthrise, the Earth seems bigger than it really is. When I saw the Earthrise come up, I said, 'Hey, now's a good time to take pictures.' So everybody grabbed for a camera, but it was Anders's camera that got it, after I had definitely told him how to pose the picture."

"Actually, "Lovell added, "this is very similar to the Iwo Jima flag-raising picture. They took a bunch of pictures there. We took a bunch of pictures of the Earth."

A lasting legacy of the Apollo photos is that they allowed people around the world to share in one of humankind's crowning achievements in a very personal way. Some contend that the photos' most enduring accomplishment was not what they revealed about the moon, but what they showed us about ourselves and our place in the cosmos.

"Congress had some hearings in the early 1970s with a lot of intellectuals and journalists and authors and thinkers about what Apollo meant, and [former *Saturday Review of Literature* editor] Norman Cousins's reply I think was pretty good," Scott recalled. "He said the significance of Apollo was not so much that man set foot on the moon, but that he set eye on the Earth. And I think all the pictures we brought back of the Earth might have been just as significant as those we took of the moon."

The photographs in this book capture the epic journey of Apollo as seen through the eyes of the explorers who helped make history and brought a piece of it back for all of the world to share.

PREVIOUS SPREAD (LEFT)
Apollo 12 astronaut Alan L. Bean holds a special environmental sample container filled with lunar soil collected during his Extravehicular Activity (EVA). A Hasselblad camera is mounted on the chest of his spacesuit. Charles P. "Pete" Conrad Jr., who took this picture, is reflected in Bean's helmet visor, November 20, 1969.

OPPOSITE
One of the goals of the Apollo 12 mission was to visit the *Surveyor 3* spacecraft that had landed in the lunar Ocean of Storms in 1967. Here Bean simulates a photographic inspection of *Surveyor 3* during training.

OVERLEAF
This section of a panoramic photograph consisting of twenty-seven separate frames taken by Charles M. Duke Jr. shows the Apollo 16 landing site in the lunar highlands, April 23, 1972. Learning how to take the panoramic photos required meticulous training and practice.

MOM
TALK

UNITED
STATES

NASA
NASA
D. F. EISELE
CUNNINGHAM

APOLLO 7

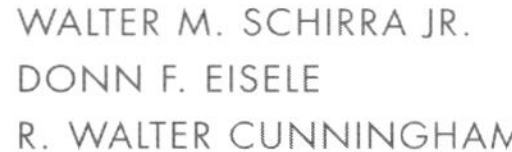

WALTER M. SCHIRRA JR.
DONN F. EISELE
R. WALTER CUNNINGHAM

Just twenty-one months after the Apollo 1 tragedy, Apollo 7 lifted off for its shakedown flight from the Kennedy Space Center. Launch occurred October 11, 1968. This was the first human flight of the new Apollo spacecraft. The main objectives of the mission were to test the many systems of the vehicle's command and service modules, including test firing the engines that would place the astronauts in lunar orbit.

Because the mission was not carrying a lunar module, Apollo 7 employed a Saturn 1B booster instead of the larger Saturn V rocket used for every other moon-related Apollo launch. In addition to being the first flight of the Apollo spacecraft, the mission was the first flight of the new Apollo spacesuits and featured the first live television broadcast from orbit.

Commander Wally Schirra had announced his retirement from NASA before the flight. He was forty-five years old and became the oldest astronaut to fly in space. He was the first to fly on Mercury, Gemini, and Apollo. The eleven-day mission took its toll on Schirra and crewmates Donn Eisele and Walter Cunningham. All developed colds and often challenged mission control over the flight's ambitious work schedule. However, the crew and spacecraft performed well and the Earth-orbital flight proved the space-worthiness of the new Apollo vehicle.

The Apollo 7 mission carried one 70-mm camera. There were a total of 533 exposures made on nine magazines of film; 498 images on color film and thirty-five images on black-and-white film.

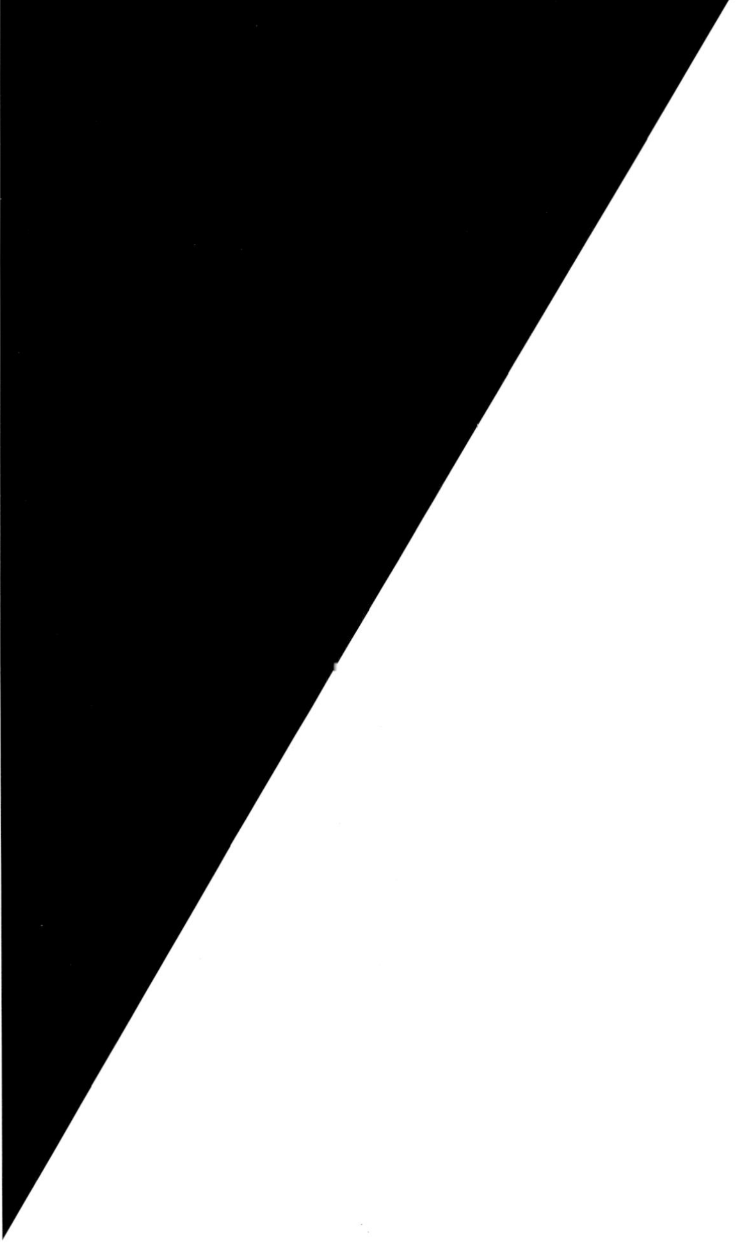

OPPOSITE
Pausing at the doorway of the recovery helicopter at the end of their mission, the Apollo 7 crew boards the USS *Essex* on October 22, 1968 (from left to right: Walter Schirra, Donn Eisele, and Walter Cunningham).

LEFT
Apollo 7 astronauts Walter Schirra (right) and Donn Eisele are seen in the very first live television transmission from space on October 14, 1968. Schirra displays a sign that reads, KEEP THOSE CARDS AND LETTERS COMING IN, FOLKS!

ABOVE
Walter Schirra performs mission duties during the Apollo 7 mission in October 1968.

OPPOSITE
The expended Saturn IVB stage is being used as a target for simulated docking maneuvers over Sonora, Mexico, during Apollo 7's second revolution around Earth on October 11, 1968.

OVERLEAF (LEFT)
The Tuamotu Archipelago in the South Pacific Ocean (looking southwest) can be seen during the Apollo 7 crew's 141st revolution around Earth on October 20, 1968.

OVERLEAF (RIGHT)
Hurricane Gladys rages in the Gulf of Mexico during the Apollo crew's ninety-first revolution around Earth on October 17, 1968.

OPPOSITE
The morning sun reflects on the Gulf of Mexico and the Atlantic Ocean as seen from the Apollo 7 spacecraft during its 134th revolution of the Earth on October 20, 1968.

This early morning picture was taken at 0646 (local time) on our ninth day in space. We passed that way each day but, due to cloud cover from Hurricane Gladys or all of the windows pointing toward outer space, we did not see this view. Everything came together on day nine, and we found ourselves looking at the Florida Peninsula, which had been our home for much of the preceding three years. Grabbing the Hasselblad camera, I perpetrated a photographic no-no, taking this picture looking into the sun.

With the sun just 9 degrees above the horizon, the picture is either enhanced or spoiled by reflections off the window and the Gulf of Mexico, depending on how you look at it. With just a little bit of effort we can see Tampa and St. Petersburg on the near coast, Orlando in the center of the state, and then Cape Canaveral and Kennedy Space Center, from which Apollo 7 was launched, on the far side. A little farther south we see Lake Okeechobee and, with a little imagination, Miami.

—R. WALTER CUNNINGHAM

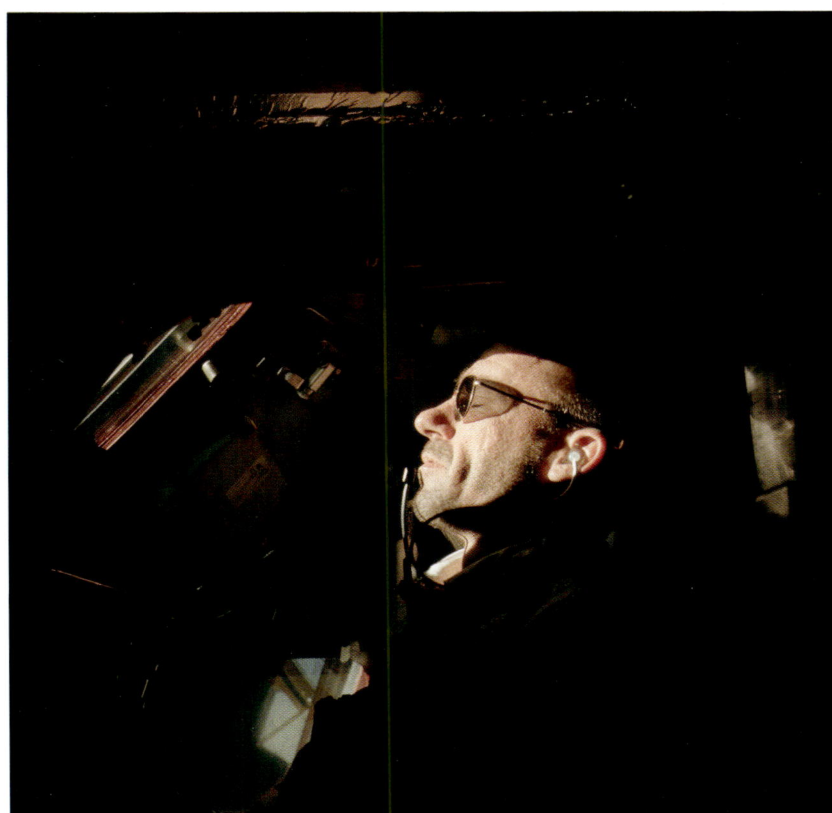

RIGHT
Wearing sunglasses to protect his eyes, astronaut Walter Cunningham is photographed during the Apollo 7 mission.

BALL

APOLLO 8

FRANK BORMAN
JAMES A. LOVELL JR.
WILLIAM A. ANDERS

The Apollo 8 mission was supposed to be the first checkout flight for the new lunar lander, but technical delays and rumors of a Soviet Union moon mission pushed NASA to make a bold decision. Frank Borman, Jim Lovell, and William Anders would become the first humans to venture beyond Earth orbit and visit another world.

The first launch of the giant Saturn V booster with people atop it came December 21, 1968, from the Kennedy Space Center. The crew departed Earth at 10:17 AM, when mission control gave the astronauts a "go" for the firing of the third-stage rocket that propelled them to the moon.

The journey took three days. Apollo 8's crew entered lunar orbit on Christmas Eve and became the first humans to see the moon's far side. For the next twenty hours, the crew studied a landscape that appeared to be a lifeless wasteland made up of gray sand and craters. They scouted future landing sites, took photographs, and used their cameras to capture the first Earthrise as seen from the moon. During the first live television broadcast from the moon, the astronauts read Bible passages from the Book of Genesis.

On Christmas Day, 1968, during the tenth orbit around the moon, the crew fired the spacecraft's main engine to return to Earth. The mission's success gave the entire Apollo program an important boost and secured a lead in the 1960s space race that America would not relinquish.

The Apollo 8 mission carried two 70-mm cameras. There were a total of 865 exposures made on seven magazines of film. The astronauts captured 589 images on black-and-white film and 276 on color film.

OPPOSITE
On December 30, 1968, the Apollo 8 crew participates in a technical debriefing session (from left to right: James Lovell, Frank Borman, and William Anders).

OVERLEAF
A panoramic view of the lunar surface includes the dark-floored Lomonosov Crater (lower right) and the bright-rayed Giordano Bruno Crater. Photograph by William A. Anders, taken as the spacecraft was leaving lunar orbit.

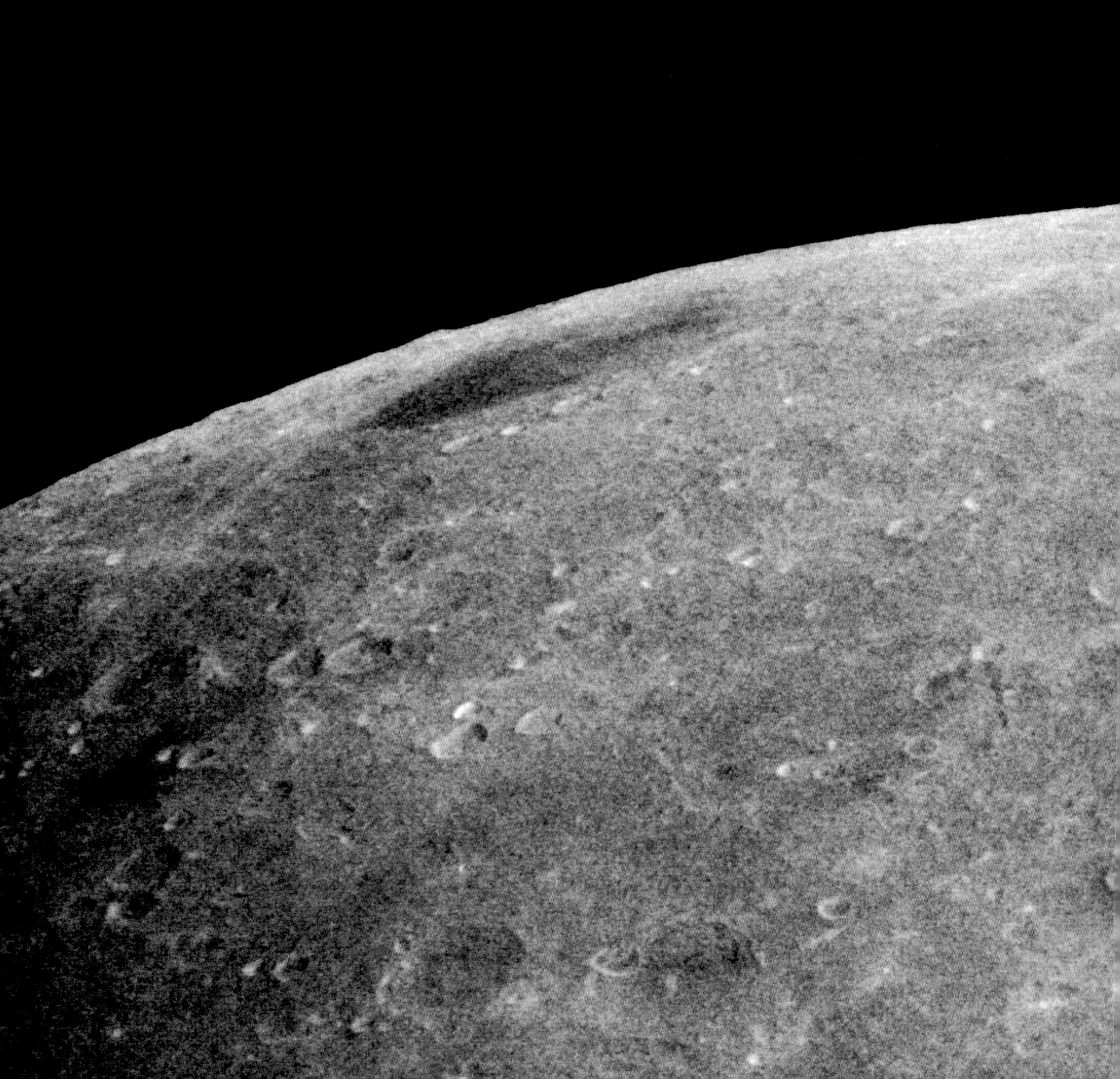

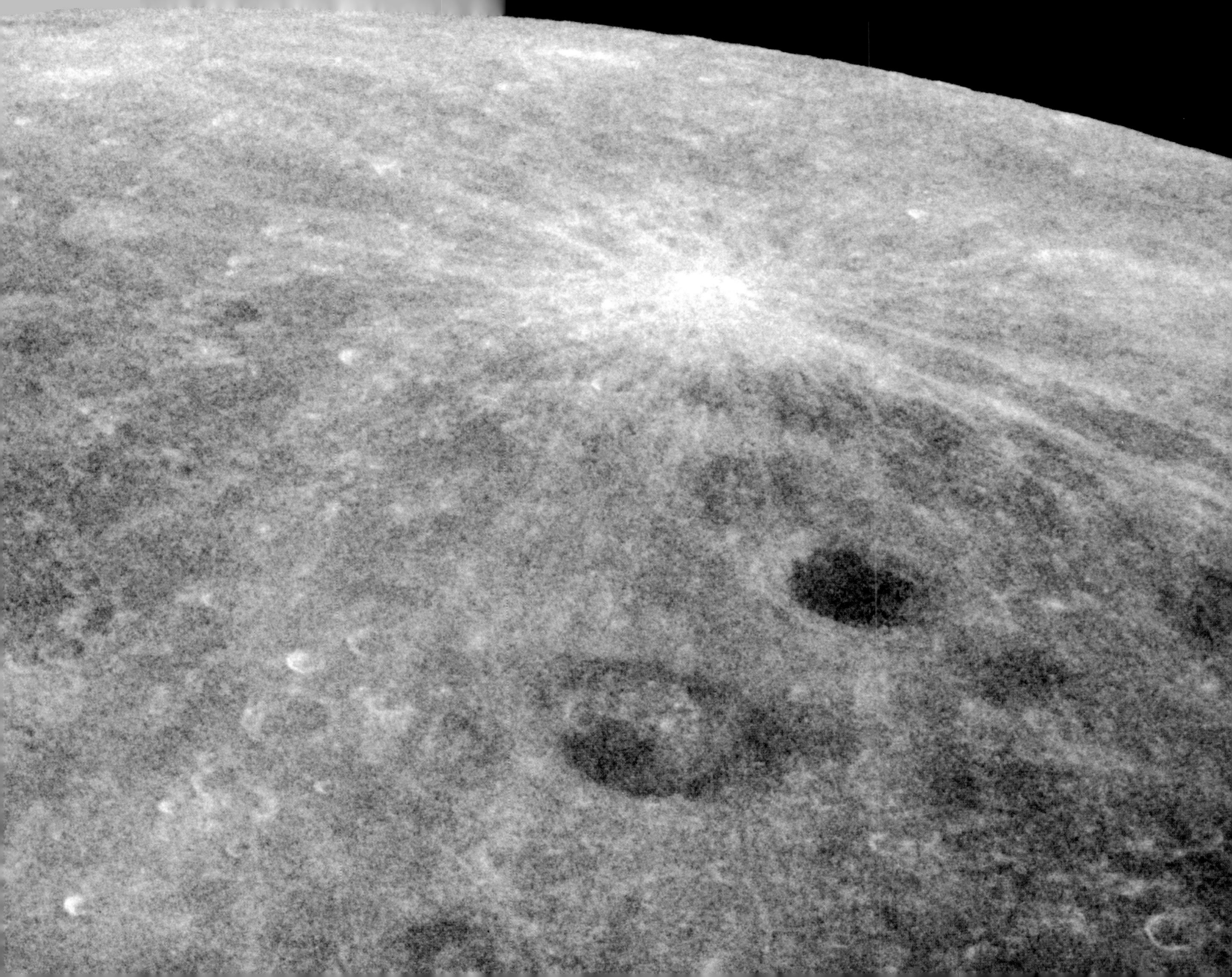

OPPOSITE
View of rising Earth about five degrees above the lunar horizon, December 22, 1968.

■ Along with my primary duties as the lunar module pilot (actually flight engineer and copilot would have been more accurate, since we had no lunar module), I was also the designated photographer for this first manned mission to leave the Earth. Consequently, I spent some time getting briefed by real photographers to hone my skills and become familiar with the onboard cameras. As I recollect, we had three 70-mm Hasselblad reflex cameras with assorted lenses and multiple magazines of 200-exposure thin-base film with various light sensitivities for different light conditions. In addition, we had a 16-mm movie camera with mountings for a window or the navigation system's optics.

To my continued amazement, we had no light meter, since f-stops and exposure times had been pre-calculated for the spacecraft's ever-changing position above the moon (and therefore reflected sunlight intensities) and the very detailed photographic target plan I was charged to execute. There was nothing in the plan for an Earthrise photo. Indeed, we didn't even see an actual Earthrise until, on our third orbit, we changed the spacecraft's orientation to heads up and looking forward. As we came around the back side of the moon, where I had been taking pictures of craters near our orbital track, I looked up and saw the startlingly beautiful sight of our home planet "rising" up above the stark and battered lunar horizon. It was the only color against the deep blackness of space. In short, it was beautiful, and clearly delicate. Frank Borman called for a camera but, since he did not like the 250-mm lens I had been using with color film, I quickly fitted one of the other cameras with a shorter lens and floated it over to him. Jim Lovell also grabbed a short-lens camera and we all blazed away at this astounding vision. There is still some debate as to which of us took the very first shot, but, regardless of who was first, the happy combination of a long lens, color film, my varying f-stops as I shot, and a much cleaner window on my side of the spacecraft, all resulted in one of the pictures from my camera's magazine being selected by NASA as what has since become the iconic "Earthrise." This picture has been on postage stamps; helped kick-start the environmental movement; Al Gore used it to introduce *An Inconvenient Truth*, his Academy-Award-winning film; and it has been replicated innumerable times. It is frequently cited as one of the most influential photos of the twentieth century—helping all here on Earth to better view our home planet as not only beautiful but also small, delicate, and worth preserving for future generations.

I'm proud to be the photographer credited with "Earthrise"—but consider it an Apollo 8 team effort.

—WILLIAM A. ANDERS

■ This one photograph brought forth our whole reason for being. We saw for the first time a fragile planet orbiting a normal star, tucked away in the outer edge of a galaxy; the Milky Way—only one of millions of galaxies in the universe. It confirmed our suspicions on how lucky we are to have a life-sustaining planet to return to. And it made us realize we had a home with limited resources and its inhabitants must learn to live and work together.

—JAMES A. LOVELL JR.

LEFT
The Apollo 8 crew is seen inside Apollo Boilerplate 1102A during water egress training in the Gulf of Mexico on October 25, 1968 (from left to right: William Anders, James Lovell, and Frank Borman).

OPPOSITE
A striking view of Earth, taken by Frank Borman on the outward journey, shows nearly the entire Western Hemisphere.

We flew to the moon as path-finders for future Apollo missions. The first view of the moon was mesmerizing, as we were aware that no other humans had seen the far side of the moon directly.

The Earth, however, captured my attention. It was the only object in the universe that we could see that had color. It was beautiful—blue with white clouds—serene, and majestic. It was home.

—FRANK BORMAN

OVERLEAF
More than two thousand people welcome the members of the Apollo 8 crew (center, in front of the microphones) home on December 29, 1968, at Ellington Air Force Base in Texas.

WELCOME

APOLLO 9

JAMES A. McDIVITT
DAVID R. SCOTT
RUSSELL L. SCHWEICKART

Apollo 9 was the first space test for the final piece of the Apollo moon landing hardware—the lunar module. It was the first vehicle designed to operate only in the vacuum of space. The lunar module was relatively light and had no heat shield like all previous human spacecraft.

For the first time since Gemini 3 in 1965, the crew was allowed to give the spacecraft names. Astronauts James McDivitt, Dave Scott, and Rusty Schweickart labeled their leggy lunar module *Spider* and dubbed the Apollo command module *Gumdrop*.

The crew launched into Earth orbit on March 3, 1969. The mission was designed to put both spacecraft through their paces, flying together as one unit, then separately to test rendezvous and docking capabilities. This flight was also the first to test, in the vacuum of space, the backpacks that would serve as portable life-support systems for astronauts on the lunar surface.

During the course of the ten-day mission, the astronauts docked and undocked their spacecraft and test-fired their engines to simulate the journey to the moon's surface. McDivitt and Schweickart flew the Apollo lunar module, while Scott remained inside the command module.

During a pivotal spacewalk, Schweickart successfully climbed out on the lunar module's front platform as Scott observed from the open hatch of the nearby command module. The exercise tested equipment that would be used to walk on the moon.

After 151 Earth orbits, the Apollo 9 crew splashed down safely in the Atlantic Ocean. The successful mission set the stage for a final dress rehearsal before NASA would attempt a landing on the moon.

The astronauts on the Apollo 9 mission carried six 70-mm cameras into space, some of which were used for a special experiment. The number of images taken on eleven magazines of film was 318 in black-and-white and 787 in color.

OPPOSITE
Fish-eye lens view of the interior of the Apollo Lunar Module Mission Simulator at Kennedy Space Center during Apollo 9 simulation training on February 23, 1969. James McDivitt is in the foreground; Russell Schweickart in the background.

OPPOSITE
The lunar module awaits extraction from Apollo 9's Saturn IVB stage, as seen from the command module.

ABOVE
James McDivitt inside the command module, March 6, 1969.

LEFT
Schweickart stands on the lunar module porch during his spacewalk on the fourth day of the flight. This photograph was taken from inside the lunar module by James McDivitt, March 6, 1969.

OPPOSITE
David Scott emerges from the hatch of the command module as it is docked with the lunar module, March 6, 1969. Schweickart took this photograph from the front porch of the lunar module.

■ I took this shot of Dave Scott (taking a picture of me!) at the beginning of my EVA on Apollo 9. It captures just a bit of the fantastic beauty of the Earth juxtaposed against the infinite black of space. In the foreground is that amazing combination of human and machine that is enabling us to emerge into the universe . . . out of the womb of Earth.

—RUSSELL L. SCHWEICKART

■ The first Apollo EVA—I've prepared the command module (*Gumdrop*) to receive Rusty Schweickart during an EVA transfer from the lunar module (*Spider*) as Jim McDivitt leads Apollo 9 through the connoisseur's test flight of the full-up Apollo mission architecture, Apollo 9, March 3–13, 1969.

—DAVID R. SCOTT

LEFT
The Apollo 9 command and service modules as seen from a window in the lunar module.

OPPOSITE
With its landing gear deployed, the lunar module *Spider* can be seen with its surface sensors extending from the footpads on March 7, 1969.

■ *Spider* was the first lunar module to fly with men aboard. It worked very well, got us safely back to the command module, and cleared the spacecraft for the flights to the moon. We were all proud of the success and fired up to proceed to the moon.

—JAMES A. McDIVITT

OPPOSITE
Apollo 10 astronauts Thomas Stafford, Eugene Cernan, and John Young are shown on the recovery ship's telephone as they receive a call from President R chard M. Nixon. This photograph was taken after the successful splashdown of the Apollo capsule.

APOLLO 10

EUGENE A. CERNAN
JOHN W. YOUNG
THOMAS P. STAFFORD

Although the mission was publicly labeled only a dress rehearsal, the flight of Apollo 10 was the second time astronauts orbited the moon and the first flight test of the lunar module above the surface. It was a challenging mission that fully tested all of Apollo's components.

Apollo 10 roared into space May 18, 1969. It was the fourth crewed Apollo flight in seven months. Because the mission required the lunar module to skim the moon's surface and snoop around, astronauts Gene Cernan, John Young, and Thomas Stafford named the vehicle *Snoopy*. Appropriately, the Apollo command module was labeled *Charlie Brown*.

On May 21, 1969, both spacecraft entered lunar orbit. Astronauts Cernan and Stafford undocked *Snoopy* and began their lunar approach, leaving Young aboard *Charlie Brown*. An hour later, the lunar module descended to within ten miles of the moon's surface. There were some tense moments during the decent, when a faulty switch setting caused the lunar module to gyrate wildly. Stafford and Cernan regained control just seconds before disaster. They finished their survey of the Apollo 11 landing site at the moon's Sea of Tranquility and safely returned to the command module.

The crew arrived back on Earth May 26, 1969, after thirty-one orbits of the moon. Among its firsts, Apollo 10 broadcast more than five hours of live color television programming from space.

The Apollo 10 mission carried two 70-mm cameras. There were a total of 1,436 exposures made on nine magazines of film, including 1,021 images in black- and-white and 415 in color.

YOUNG

OPPOSITE
The Apollo 10 astronauts undergo water-egress training in a tank at the Manned Spacecraft Center (MSC) in Houston in August 1968. Leaving the command module trainer is commander Stafford. Cernan (in foreground) and Young are already in the raft.

ABOVE
Replicas of Snoopy and Charlie Brown, two characters from Charles Schulz's famous comic strip, *Peanuts*, decorate the top of an operations console in the Mission Control Center, Houston, on the first day of the Apollo 10 lunar orbit mission.

LEFT
Apollo 10 surveyed proposed landing sites for Apollo 11. This view by John Young shows one of them in the Central Bay area of the moon. Bruce, the prominent crater near the bottom of the scene, is about six kilometers (3.7 statute miles) in diameter.

OPPOSITE
The Apollo 10 command module *Charlie Brown* as seen from the lunar module *Snoopy* after separation in lunar orbit. Numerous bright craters and the absence of shadows show that the sun was almost directly overhead when this photograph was taken.

■ Apollo 10 command and service module, circling the moon. In the command module I am flying solo while Tom Stafford or Gene Cernan took the picture with a 250-mm Hasselblad camera.

—JOHN W. YOUNG

OPPOSITE
A view of the Earth from thirty-six thousand nautical miles away taken during Apollo 10's journey toward the moon, May 18, 1969. Nearly all of Mexico north of the Isthmus of Tehuantepec can be clearly identified.

This is my favorite photo from Apollo 10. Due to the time of year, trajectory, and weather, it is one of the best photos of the North American continent from the Apollo program. To see our home planet from this view was absolutely awesome. It was nearly breathtaking. I was reminded of Socrates's saying in 399 BCE, before flight above the ground was theorized: "A man must rise above the Earth to the top of the atmosphere and beyond, and only thus will he fully understand the world in which he lives."

I could look at this great view for hours and try to capture the meaning of it all. I thank God I had the opportunity to fly this incredible mission with my fellow crew members and see the sights that only few in the history of the planet have ever seen.

—THOMAS P. STAFFORD

RIGHT
Apollo 10 crew members Cernan, Stafford, and Young step from the helicopter that plucked them out of the ocean to a red-carpet welcome aboard the USS *Princeton*, May 26, 1969.

Many folks wonder whether Apollo 10 was a disappointment because we only came close, but didn't land.

The expressions on our faces should answer that question. Besides, we painted the white line in the sky so Neil wouldn't get lost! Having had the opportunity to see our home from a quarter million miles away was an overwhelming and lasting personal experience—one I shall never forget—an experience only explained by the existence of a creator, which was reinforced during my second voyage to the moon on Apollo 17.

—EUGENE A. CERNAN

NASA
NASA

APOLLO 11

NEIL A. ARMSTRONG
BUZZ ALDRIN
MICHAEL COLLINS

"Men have landed and walked on the moon." That was how the *New York Times* recorded the events of July 20, 1969, when NASA met President Kennedy's challenge and pulled off one of the greatest achievements in human history. On that day, American astronauts set foot on another celestial body.

The straightforward but monumental goal of Apollo 11 was to "perform a manned lunar landing and return." The mission began the morning of July 16, 1969, with astronauts Neil Armstrong, Buzz Aldrin, and Michael Collins seated in the command module *Columbia* atop a Saturn V rocket at the Kennedy Space Center. At 9:32 AM EDT, the three-stage, thirty-six-story rocket used its 7.5 million pounds of thrust to propel the crew into space—and history.

Three days later, Armstrong and Aldrin climbed into the lunar module to attempt the first-ever landing on the moon's surface. Their spacecraft was named *Eagle* in honor of the winged symbol of America, which was also the inspiration for the flight's mission insignia. Collins, who stayed behind aboard *Columbia*, would write later that the lunar module was "the weirdest-looking contraption I have ever seen in the sky."

During the final seconds of *Eagle's* decent, with computer alarms sounding and fuel running dangerously low, Armstrong manually took control of the ship and steered clear of boulders that littered the landing site. At 4:18 PM on July 20, 1969, Armstrong finally radioed back to an anxious Earth, "Houston, Tranquility Base here. The *Eagle* has landed."

Six hours later, at 9:56 PM, Armstrong took his "one giant leap for mankind." Aldrin joined him on the moon about fifteen minutes later and the astronauts spent two and a half hours collecting soil samples and rocks, setting up experiments, and photographing the surface from a vantage point considered impossible less than a decade earlier. The crew completed a triumphant return to Earth four days later with a splashdown in the Pacific Ocean.

The astronauts on Apollo 11 carried four 70-mm cameras to the moon. There were a total of 1,407 exposures made on nine magazines of film. They captured 857 images on black-and-white film and 550 on color film.

PREVIOUS SPREAD (LEFT)
The Apollo 11 crew relaxes on the deck of the NASA Motor Vessel Retriever prior to water-egress training in the Gulf of Mexico, May 24, 1969. Left to right are Buzz Aldrin, Neil Armstrong, and Michael Collins. In the background is Apollo Boilerplate 1102, used in the training exercise.

OPPOSITE
Buzz Aldrin steps onto the lunar surface during the Apollo 11 EVA on the moon, July 20, 1969. Armstrong took this photograph with a 70-mm lunar surface camera he shared with Aldrin.

ABOVE
In this photograph of the lunar module, Armstrong's shadow can been seen with the shadow of the Gold camera, used for close-up stereoscopic photography.

OPPOSITE (TOP LEFT)
A view from the north of the lunar module. Photograph taken by Buzz Aldrin.

OPPOSITE (TOP RIGHT)
Aldrin has placed the Laser Ranging Retro-Reflector (LRRR), on the surface and has turned to his left to find a spot for the seismometer. Photograph by Neil A. Armstrong.

OPPOSITE (BOTTOM LEFT)
Aldrin deploys the passive seismic experiments package. Photograph by Neil A. Armstrong.

OPPOSITE (BOTTOM RIGHT)
Aldrin is in the process of taking a lunar soil sample. In the background is the deployed solar wind experiment. Both Aldrin and Armstrong spent approximately two hours and twenty minutes walking on the lunar surface. Photograph by Neil A. Armstrong.

LEFT
Aldrin poses for one of the best-known photographs taken during the Apollo 11 mission. He is standing beside the deployed U.S. flag. The lunar module *Eagle* is on the left and footprints of the astronauts are clearly visible in the lunar soil.

OVERLEAF
Apollo astronauts trained on Earth to take individual photographs in succession in order to create a series of frames that could be assembled into panoramic images. This frame from Aldrin's panorama of the Apollo 11 landing site is the only good picture of Armstrong on the lunar surface.

UNITED
STATES

OPPOSITE
Aldrin walks on the moon near the leg of the lunar module *Eagle*.

■ Nothing prepared me for the starkness of the moon. The barren terrain was a dusty gray with many little craters in every direction. The sky was utter blackness, void of any stars. When I stepped down onto the surface and felt each movement carried by the slow-motion sensation of one-sixth lunar gravity, I spontaneously exclaimed, "Magnificent desolation." As I walked away from the *Eagle* lunar module, Neil said, "Hold it, Buzz." So I stopped and turned around, and then he took what has become known as the "Visor" photo.

I like this photo because it captures the moment of a solitary human figure against the horizon of the moon, along with a reflection in my helmet's visor of our home away from home, the *Eagle*, and of Neil snapping the photo. Here we were, farther away from the rest of humanity than any two humans had ever ventured. Yet in another sense we became inextricably connected to the hundreds of millions watching us more than 240,000 miles away. In this one moment the world came together in peace for all mankind.

–BUZZ ALDRIN

RIGHT
The now-famous close-up view of Aldrin's bootprint in the lunar soil. Aldrin made this footprint on the pristine surface of the moon so he could then photograph it for study by soil mechanics experts.

OPPOSITE
The *Eagle*'s ascent stage, with astronauts Armstrong and Aldrin aboard, is photographed by Collins in the command module *Columbia* during rendezvous in lunar orbit, July 21, 1969. The large, dark-colored area in the background is Smyth's Sea. Earth rises above the lunar horizon.

■ Of the various uncertainties during the flight of Apollo 11, returning Neil and Buzz in one piece to the command module was paramount in my mind . . . . Just prior to my taking this photograph, they had departed Tranquility Base using the lower half of their landing craft, *Eagle*, as a launch platform. The upper half (second stage) appeared first as a tiny gold insect crawling across the lunar landscape, and then began to take form as a man-made object, although its angular shape still seemed strange and awkward to me. Little by little, they grew closer, steady, as if on rails, and I thought, "What a beautiful sight," one that had to be recorded. As I reached for my Hasselblad, suddenly the Earth popped up over the horizon, directly behind *Eagle*.

I could not have staged it any better, but the alignment was not of my doing, just a happy coincidence. I suspect a lot of good photography is like that, some serendipitous happenstance beyond the control of the photographer. But at any rate, as I clicked away, I realized that for the first time, in one frame, appeared three billion earthlings, two explorers, and one moon. The photographer, of course, was discreetly out of view.

—MICHAEL COLLINS

■ This picture captures the essence of the Apollo project, two celestial bodies connected by a human transportation system composed of two very different vehicles. Taken through the window (window frame in the lower left) of the command module which has the ability to return though the Earth's atmosphere at very high speeds, this photograph features the ascent stage of the lunar module, the only craft constructed to be able to convey crewmen from lunar orbit to the surface of the moon and return them to orbit around the moon. As the lunar module was unable to return through the Earth's atmosphere, it was required to rendezvous and dock with the command module, where the crew could transfer to the command module for the return to their home planet.

—NEIL A. ARMSTRONG

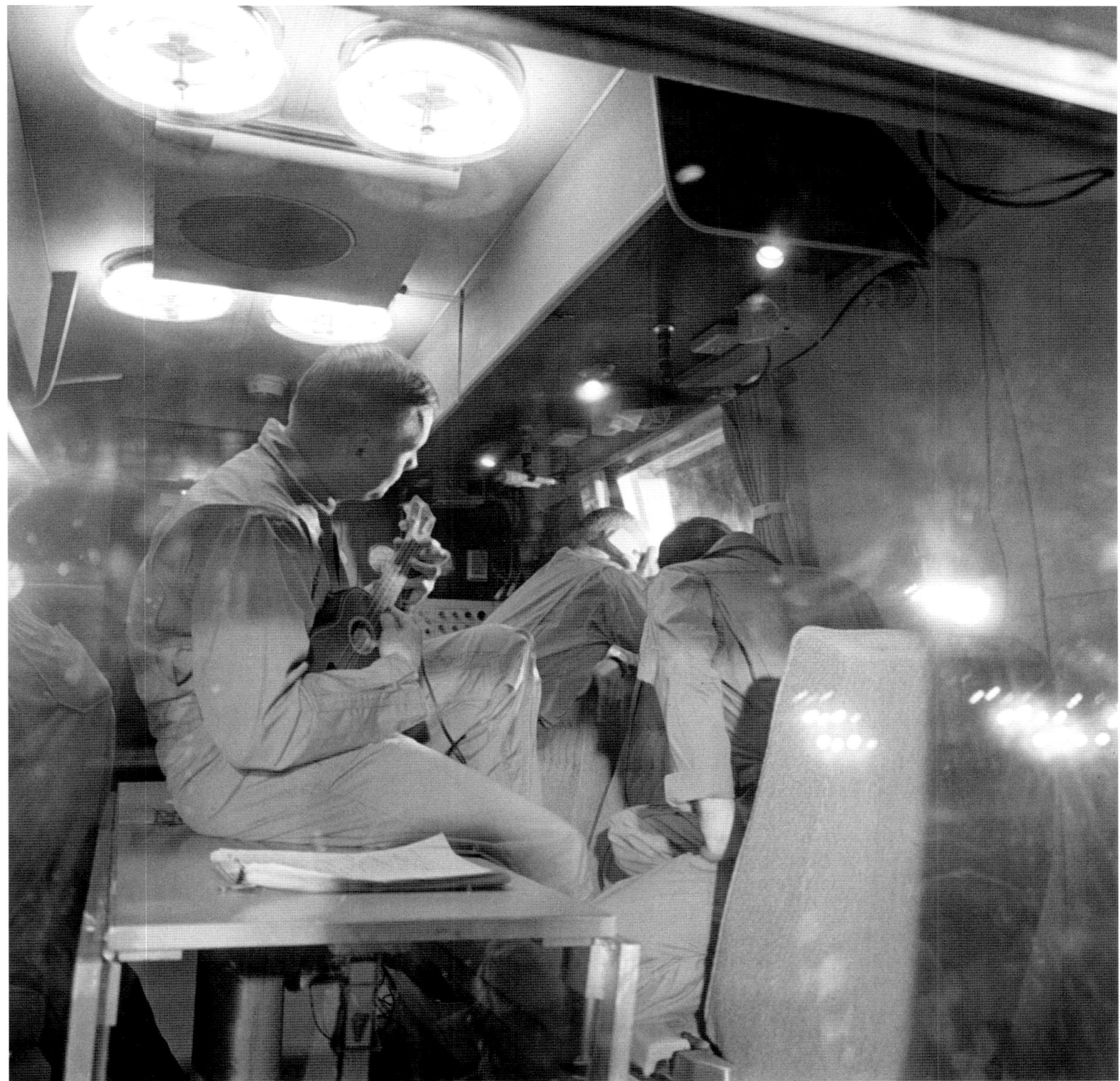

OPPOSITE
Apollo 11 capsule is photographed being lowered to the deck of the USS *Hornet*, the prime recovery ship for the mission. Note the flotation ring attached by Navy divers has been removed from *Columbia*.

ABOVE
Interior view of the Mobile Quarantine Facility, showing the crew after arrival at Ellington Air Force Base in Texas very early Sunday, July 27, 1969, after a long flight from Hawaii. A large crowd was present to welcome the heroic astronauts home. Armstrong is strumming on a ukulele.

OVERLEAF
New York City welcomes the Apollo 11 astronauts in a shower of ticker tape down Broadway and Park Avenue in a parade described as one of the largest in the city's history, August 13, 1969.

THREE MORE
LIKE BEFORE
USS HORNET CVS
12
APOLLO
AUTHORIZED PERSONNEL ONLY
U. S. GOVERNMENT QUARANTINE
STATION

APOLLO 12

CHARLES P. "PETE" CONRAD JR.
RICHARD F. GORDON
ALAN L. BEAN

In April 1967, the unpiloted robotic probe *Surveyor 3* landed on the moon's Ocean of Storms. Two years later, Apollo 12 attempted a pinpoint landing nearby so that moon walkers Charles "Pete" Conrad and Alan Bean could retrieve pieces of *Surveyor 3* and return them to Earth for analysis.

The mission got off to a bumpy start shortly after liftoff on November 14, 1969, when lightning struck the crew's Saturn V rocket. For a few moments, it appeared that the flight might be in jeopardy. However, quick thinking by mission control and Bean inside the command module *Yankee Clipper* restored the flow of data between the spacecraft and the ground; NASA was headed to the moon for a second landing attempt.

Because of the rocky lunar terrain at the Sea of Tranquility, Apollo 11 had missed its targeted landing site by about four miles. For Apollo 12 to be successful, Conrad and Bean had to land the lunar module *Intrepid* within walking distance of the *Surveyor 3* lander. With command module pilot Dick Gordon orbiting overhead, Conrad and Bean touched down within six hundred feet of the probe.

Conrad took his first steps on the moon at 6:44 AM on November 19, 1969. After having a little trouble making the last step from *Intrepid's* ladder, he exclaimed, "*Whoopee!* Man, that may have been a small step for Neil, but that's a long one for me." Bean joined him minutes later. The two astronauts made two moon walks and spent more than thirty-one hours on the lunar surface. They retrieved several pieces of *Surveyor 3*, including the spacecraft's camera.

For the first time, color television pictures were beamed live from the lunar surface. However, as Bean moved the camera to a new location, he accidentally pointed it directly at the sun, destroying the picture tube and ending television coverage. The astronauts would rely on the images they captured with their still cameras to document the mission. The crew completed a safe return to Earth November 24, 1969, with a splashdown in the Pacific Ocean.

Apollo 12 carried seven 70-mm cameras, including three individual cameras and a block of four that made up the lunar multispectral experiment. There were a total of 1,725 during the experiment. The number of images taken by the astronauts on fourteen magazines of film totaled 1,438 in black-and-white, 571 in color film, and 104 infrared photographs.

12

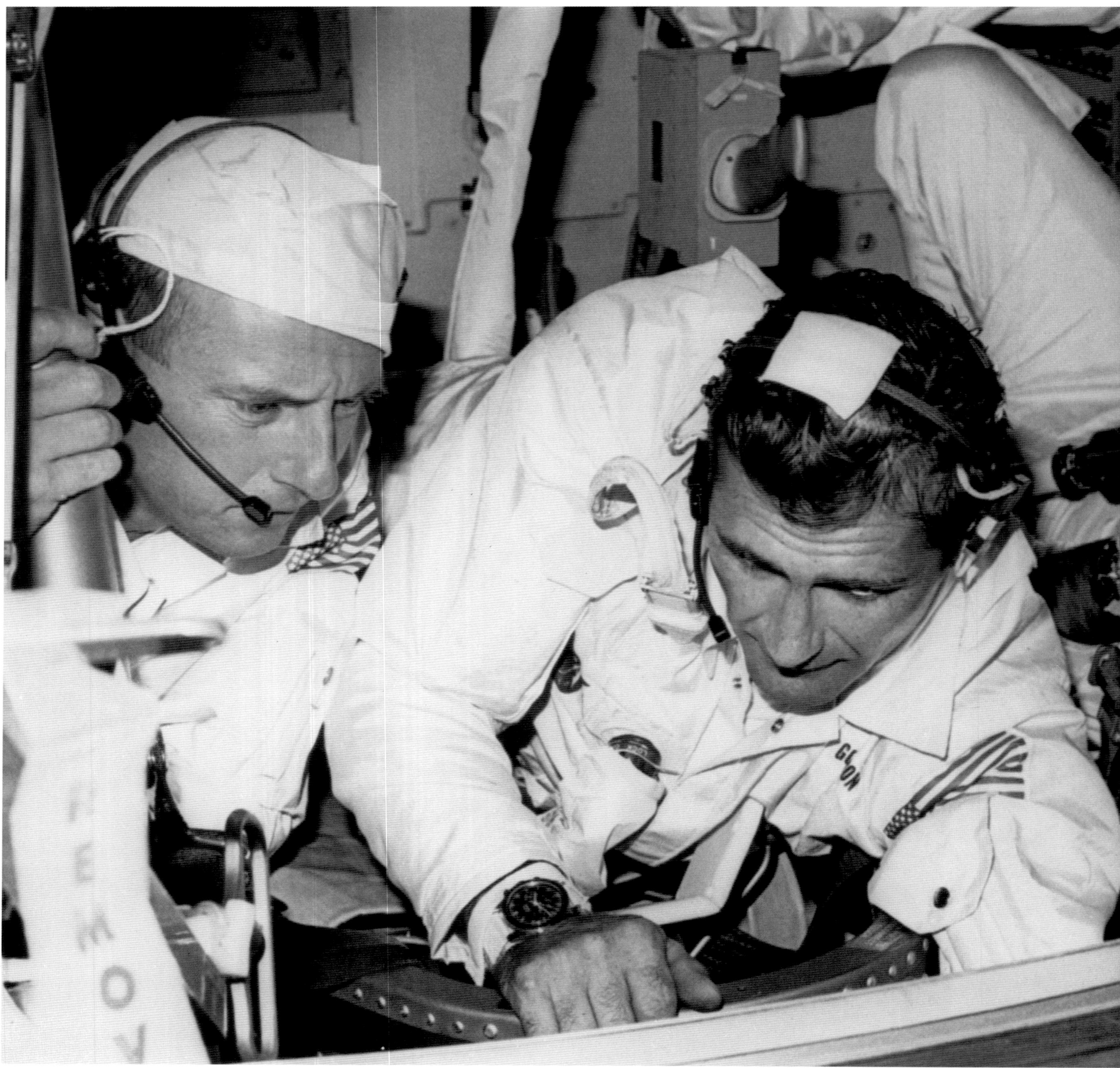

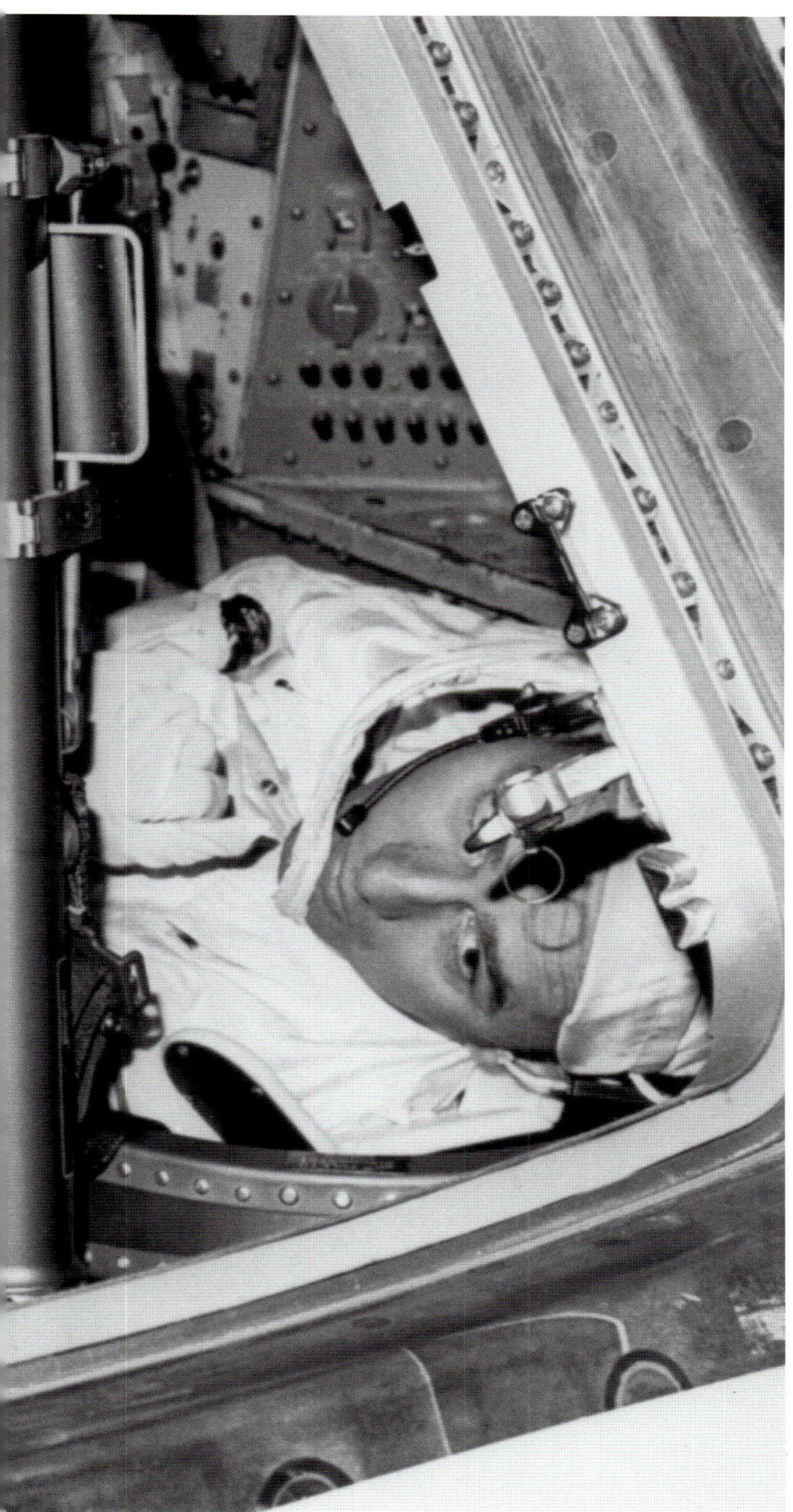

PREVIOUS SPREAD
Rear Admiral Donald C. David, commander, Manned Spacecraft Recovery Force, Pacific, welcomes the crew of Apollo 12 aboard the USS *Hornet*, November 24, 1969. The astronauts are inside the Mobile Quarantine Facility. Waving in the lead car, from the left, are Michael Collins, Buzz Aldrin, and Neil A. Armstrong.

LEFT
The Apollo 12 astronauts are photographed during spacecraft checkout activity at North American Rockwell Space Division at Downey, California. Left to right: Charles Conrad Jr., Richard F. Gordon Jr., and Alan L. Bean.

■ This is one of several of my favorite photos from my Apollo 12 mission. It might seem a little unusual, as it is of Pete Conrad, Dick Gordon, and me working together during final testing of our spacecraft just prior to its shipment to Cape Kennedy. One of the great blessings of my life was being teamed with Pete and Dick. Both were more experienced in spaceflight and life in general.They understood what was important around the Astronaut Office during and after hours, and I discovered that up to the time they took me under their wing I didn't have a clue. Pete and Dick changed my life for the better because they cared enough to go way out of their way with their wise, heartfelt advice. I think of and feel love for Pete and Dick every single day.

—ALAN L. BEAN

OPPOSITE
Charles P. "Pete" Conrad Jr. stands beside the U.S. flag after it was placed on the lunar surface during the first EVA on November 19, 1969. Photograph by Alan L. Bean.

ABOVE
Bean walks from the color lunar surface television camera (center) toward the lunar module *Intrepid*, on the moon's Ocean of Storms. Photograph by Charles P. "Pete" Conrad Jr.

OPPOSITE
This unusual photograph taken by Bean on November 20, 1969, shows two U.S. spacecraft on the surface of the moon: The Apollo 12 lunar module *Intrepid* is in the background and the unmanned *Surveyor 3* spacecraft that landed on the moon on April 20, 1967, is in the foreground, with Conrad standing beside it.

■ This image is symbolic of the mission. Our objective was to perform a pinpoint landing at the *Surveyor* 3 target site at the rim of the Surveyor Crater. This precision landing was of great significance to the future exploration program because landing points in rough terrain of great scientific interest could then be targeted. Pete Conrad is examining *Surveyor*'s TV camera prior to detaching it.

—RICHARD F. GORDON

RIGHT
Conrad stands at the Module Equipment Stowage Assembly on the lunar module, November 19, 1969. The S-band antenna is deployed at right. The carrier for the Apollo Lunar Hand Tools is near Conrad. Photograph by Alan L. Bean.

OVERLEAF
Bean is assisted in leaving the Apollo 12 command module *Yankee Clipper* by a U.S. Navy swimmer during recovery operations in the Pacific Ocean, November 24, 1969. Conrad and Gordon are already in the life raft.

315 NAVY

OPPOSITE
Astronaut secretary Martha Caballero wishes Apollo 13 commander James Lovell good luck.

ABOVE
The original Apollo 13 crew stands by for water-egress training at the Manned Spacecraft Center (MSC) in Houston on January 17, 1970. They are James A. Lovell Jr., Fred W. Haise Jr., and Thomas K. Mattingly II (in background).

JAMES A. LOVELL JR.
JOHN L. SWIGERT JR.
FRED W. HAISE JR.

Historians have concluded that some of Apollo's finest moments came during the one flight that failed to achieve its goal. Astronauts Jim Lovell, Fred Haise, and Ken Mattingly were assigned to the Apollo 13 mission, which was supposed to land in the moon's Fra Mauro region. However, the first of several bad breaks came seventy-two hours before the scheduled launch when Mattingly, the command module pilot, was removed from the mission after being exposed to German measles. Backup astronaut Jack Swigert took Mattingly's place.

Apollo 13 was launched from the Kennedy Space Center on April 11, 1970. Two days later, about two hundred thousand miles from Earth, the Apollo 13 crew completed a television broadcast from the command module *Odyssey* and the lunar module *Aquarius*. The major television networks declined to carry the transmission live because network executives thought the flights had become routine. That belief was shattered moments later when an oxygen tank in the spacecraft's service module exploded. Swigert was the first to call mission control with the often-paraphrased report, "Houston, we've had a problem here."

America's third moon landing was aborted. The new plan was simply to get the astronauts home alive. The drama unfolding in space quickly captivated the world.

With *Odyssey* running out of oxygen and power, the crew powered up *Aquarius* to serve as a lifeboat. The lunar module was designed to support two men for two days. Now, it was being asked to care for three astronauts for almost four days. Carbon dioxide rose to dangerous levels and mission managers ingeniously devised a way to attach the command module's air filters to the lunar module's system using plastic bags, cardboard, tape, and an old sock.

The ride back to Earth was cold and miserable. The crew had little food and water, and even less sleep. But the Apollo 13 mission ended safely with a splashdown in the Pacific Ocean on April 17, 1970. NASA had pulled off a feat almost as impressive as a successful moon landing.

The Apollo 13 mission carried four 70-mm cameras; however, only two were used. There were a total of 584 exposures made on five magazines of film; ninety-five images in black-and-white and 489 in color.

13

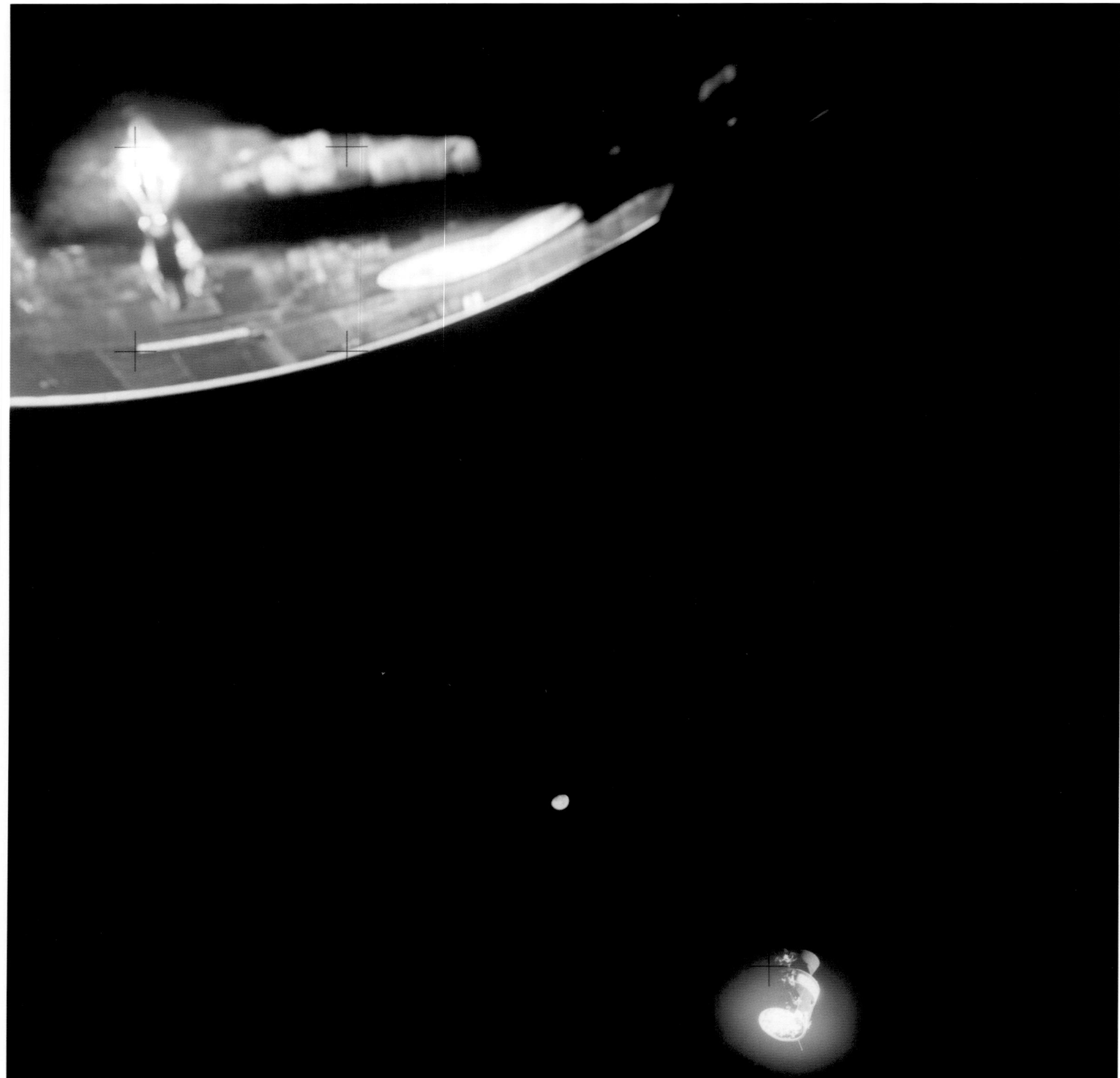

ABOVE
This view of the severely damaged Apollo 13 service module, with the moon in the distant background, was photographed from the lunar module *Aquarius* on April 17, 1970. The damage to the Apollo spacecraft forced the crewmembers to use the lunar module as a lifeboat.

OPPOSITE
An oblique view of International Astronomical Union Crater No. 302 on the far side of the moon, taken from *Aquarius*. The Apollo 13 astronauts used the gravitational force of the moon as a slingshot to send their damaged spacecraft back to Earth.

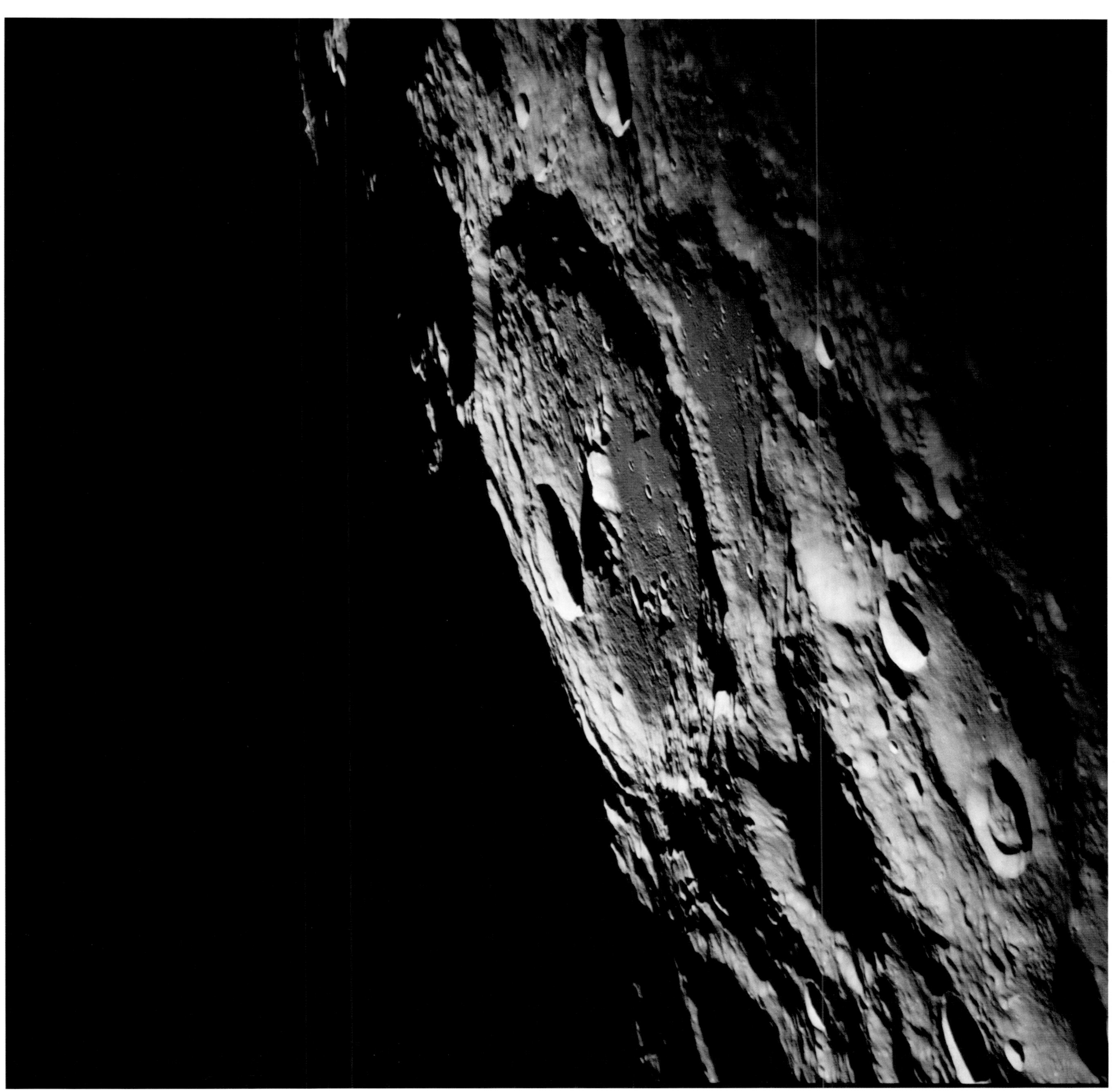

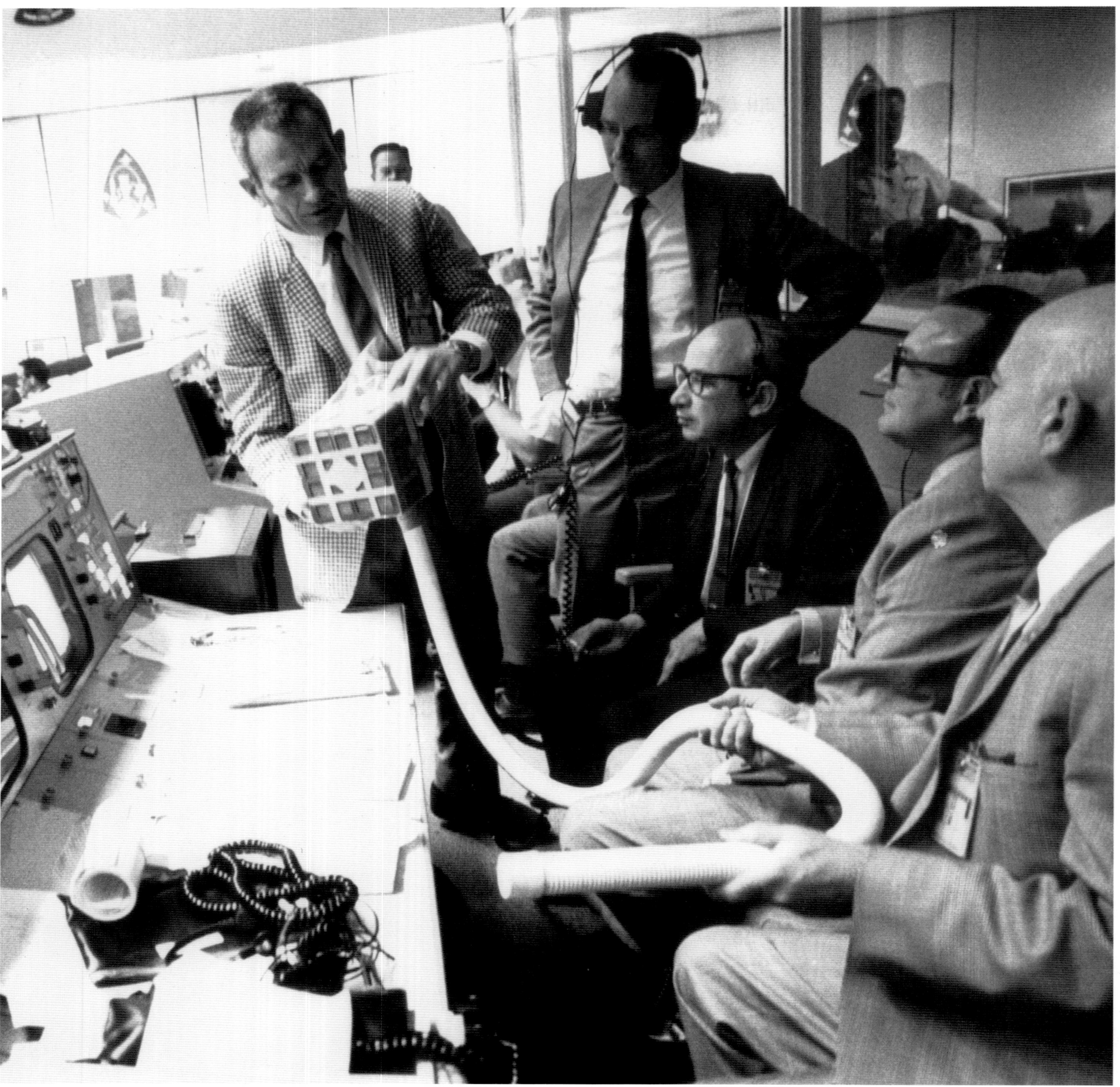

ODYSSEY
X
XI
XII
XIII
APOLLO XIII

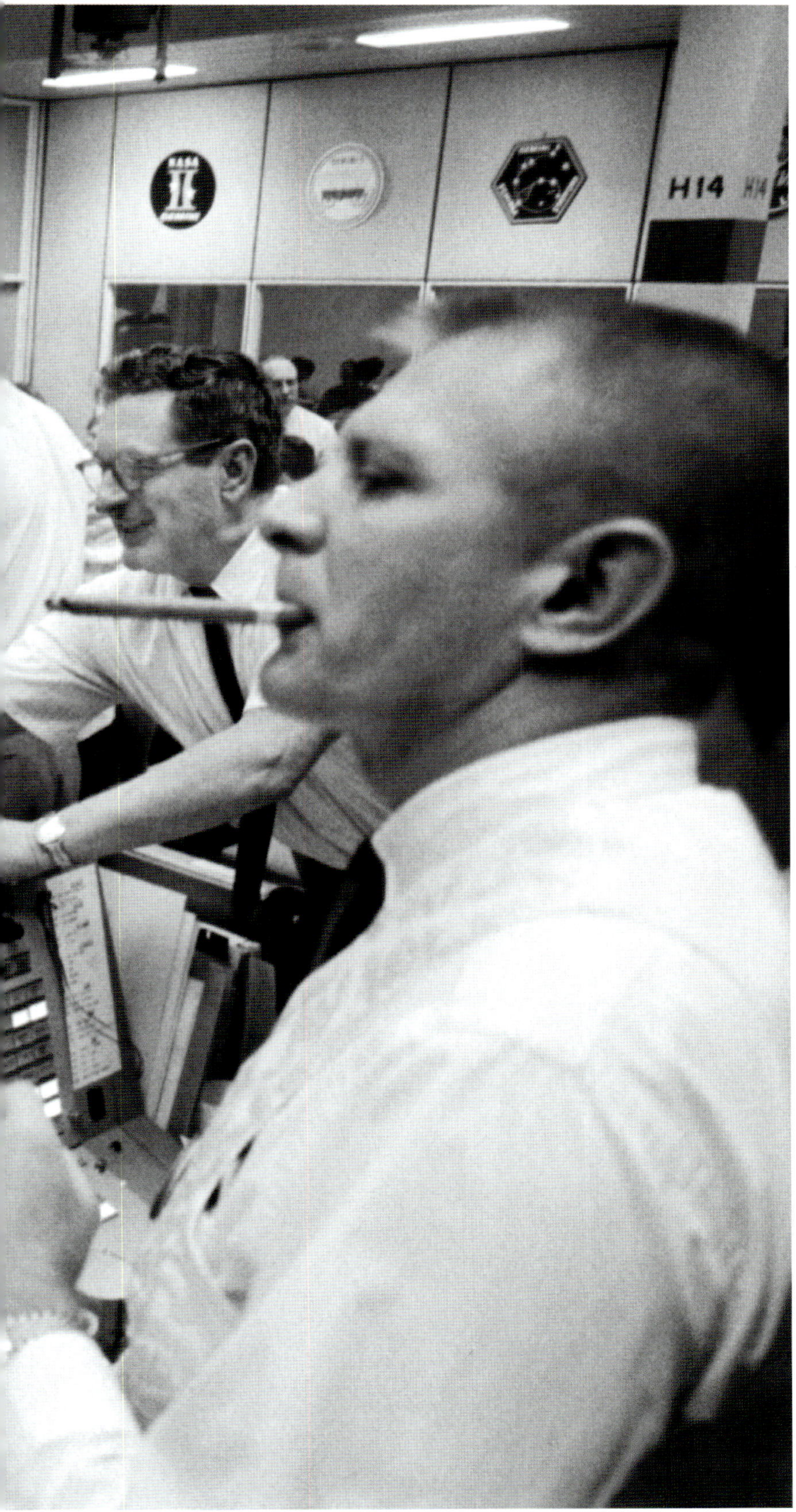

PREVIOUS SPREAD (LEFT)
On April 15, 1970, Donald "Deke" Slayton, director of flight crew operations, holds a prototype of the "mailbox" that mission control hoped to use to remove carbon dioxide from the Apollo 13 spacecraft. A spacesuit exhaust hose is connected to a lithium hydroxide canister to purge the cabin air. Looking on are (from the left): Howard Tindall, deputy director, flight operations; Sigurd Sjoberg, director, flight operations; Christopher Kraft, deputy director, MSC; and Robert Gilruth, director, MSC.

PREVIOUS SPREAD (RIGHT)
An interior view of the lunar module during the trouble-plagued journey back to Earth shows some of the temporary hose connections and apparatus that were required for the astronauts to survive in *Aquarius*. John Swigert is on the right.

LEFT
Slayton (in black shirt, left of center), director of flight crew operations, and Chester M. Lee from NASA Headquarters, Washington, shake hands in mission control, while Rocco Petrone, also from Headquarters (standing, near Lee), watches the large screen, April 17, 1970.

ABOVE
The Apollo 13 spacecraft and its parachutes are just visible against a gap in the dark clouds as they descend to a splashdown in the South Pacific Ocean, April 17, 1970.

LEFT
Fred W. Haise during training for the Apollo 13 mission.

■ This photo of myself in a training suit that shows considerable wear and tear was taken with the Saturn V to take us to the moon on Apollo 13 in the background.

For the hundreds of hours spent in water tanks, on EVA exercises on simulated lunar fields, and in mission simulators, we utilized our training space suits. Because of the less break-in time on our flight suits, they did not feel as comfortable and homey as our training ones.

The worn appearance of the suit serves well to reflect the hard work that was done by ourselves, our training staff, and those in Mission Control to assure success. The fruits of this labor in preparation for flight were born out during our Apollo 13 mission. Because of the drama, Apollo 13 stands out as an example of what the well-trained team can accomplish in overcoming major challenges. In actuality, every one of our lunar missions had problems and challenges that were overcome.

—FRED W. HAISE JR.

ABOVE
President Richard M. Nixon and the Apollo 13 crew pay honor to the U.S. flag during the post-mission ceremonies at Hickam Air Force Base, Hawaii, on April 18, 1970. Haise, Lovell, and Swigert have just been presented with the Presidential Medal of Freedom.

This photo graphically emphasizes the American absolute ability to overcome adversity. Faced with a crisis, dedicated citizens under excellent leadership that fostered teamwork, initiative, and perseverance were able to turn a certain catastrophe into a successful conclusion.

—JAMES A. LOVELL JR.

APOLLO 14

ALAN B. SHEPARD JR.
STUART A. ROOSA
EDGAR D. MITCHELL

Apollo 14 picked up the mission that was abandoned by the troubled Apollo 13 flight. The astronauts' destination was the moon's hilly region north of the Fra Mauro crater.

Commanding the mission was Alan Shepard, the first American in space and the only original Mercury astronaut to make it to the lunar surface. He had been grounded for almost five years because of an inner ear condition that caused dizziness and disorientation. Corrective surgery allowed Shepard to return to flight status in 1969. At age forty-seven, he was the oldest astronaut in the program. His Apollo 14 crewmates were Stuart Roosa, the command module pilot, and Edgar Mitchell, the lunar module pilot. The mission lifted off from the Kennedy Space Center on January 31, 1971.

The journey to the moon had its share of minor problems. Roosa had difficulty docking the command module *Kitty Hawk* to the lunar module *Antares*. It took five attempts and more than ninety minutes to connect the two spacecraft. After the two ships undocked in lunar orbit, computer and radar glitches aboard *Antares* initially threatened the descent to the surface. However, the problems were resolved, and Shepard and Mitchell safely landed their spacecraft on the moon February 5, 1971.

The astronauts made two moon walks while Roosa collected data and took photographs from *Kitty Hawk*, which was orbiting overhead. Shepard and Mitchell deployed and activated various scientific equipment and experiments, and collected almost one hundred pounds of lunar soil and rock samples. Their work on the moon was made easier by a two-wheeled cart called the Modular Equipment Transporter.

Despite Apollo 14's many technical and scientific achievements, the mission is remembered by many for Shepard's golf prowess. An avid golfer, the veteran astronaut combined the head of a six-iron and a collapsible tool handle to create a golf club. He took a couple of shots as his final moon walk was coming to an end.

The Apollo 14 mission concluded February 9, 1971, with the crew's successful splashdown in the Pacific Ocean. They had carried four 70-mm cameras to the moon and made 1,342 exposures, including 823 images in black-and-white and 519 photographs in color on fifteen magazines of film.

PREVIOUS SPREAD
The Apollo 14 crew relaxes aboard the NASA Motor Vessel Retriever, prior to participating in water-egress training in the Gulf of Mexico, October 24, 1970. Left to right are Alan Shepard Jr., Stuart Roosa, and Edgar Mitchell. They are standing by a command module trainer used in the exercises.

OPPOSITE
Alan Shepard took this photograph looking southeast toward Edgar Mitchell, who is using a device called a thumper to generate data about the internal structure of the moon, February 5, 1971. The crater beyond Mitchell is known as Old Nameless.

ABOVE
The Apollo 14 lunar module *Antares* as seen by the two moon-exploring astronauts, photographed against a brilliant sun glare during the first EVA. The bright trail leading to the lander was made by the MET.

ABOVE
Mitchell moves across the lunar surface as he looks over a traverse map. Lunar dust can be seen clinging to the boots and legs of the spacesuit. Photograph by Alan B. Shepard Jr.

■ This photograph is representative of the long trek up Cone Crater; it is one of the most sought after by collectors; it is symbolic of the difficulties we had in micro-navigation to fulfill the geologist's sample strategy, and it is representative of the pervasiveness of the lunar dust in contaminating the pressure suit.

—EDGAR D. MITCHELL

CLOCKWISE FROM RIGHT

Shepard stands beside a large boulder on the lunar surface during the mission's second EVA on February 6, 1971. Photograph by Edgar D. Mitchell.

Shepard assembles a double core tube as he stands beside the MET. Photograph by Edgar D. Mitchell.

A close-up view of the Laser Ranging Retro-Reflector, which was deployed on the moon to aid in measuring its precise distance from Earth.

A hammer and a small collection bag lie atop a lunar boulder to give some indication of size in this view of several boulders clustered together.

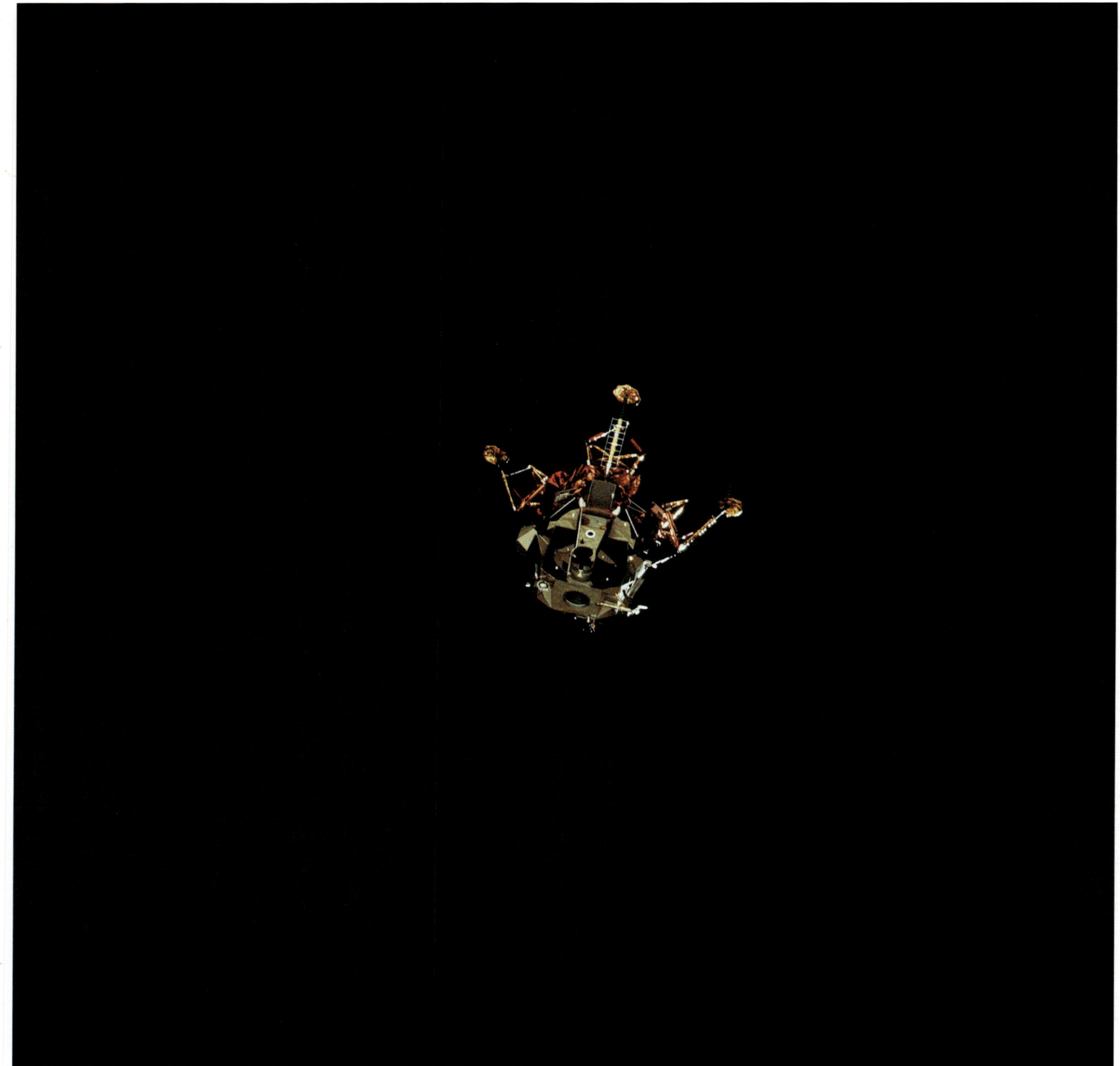

ABOVE
A view of the lunar module *Antares* taken from the command module *Kitty Hawk* by Stuart Roosa, February 6, 1971. The landing crew is safely aboard *Kitty Hawk* and the uninhabited *Antares* drifts into space.

OPPOSITE
The Apollo 14 command module descends into the South Pacific Ocean on February 9, 1971.

RESCUE

APOLLO 15

DAVID R. SCOTT
JAMES B. IRWIN
ALFRED M. WORDEN

Beginning with Apollo 15, NASA committed to more advanced lunar exploration. The program's final three missions conducted more detailed scientific studies of the moon on the surface and from lunar orbit.

Astronauts Dave Scott, Jim Irwin, and Al Worden named their command module *Endeavour* in honor of the eighteenth-century sailing ship Captain James Cook used for the first extensive science voyages. Cook likely would have envied the Apollo 15 expedition, which was designed to conduct lunar exploration over longer periods, greater distances, and with more scientific instruments than any other Apollo mission.

Apollo 15 launched from the Kennedy Space Center on July 26, 1971. Astronauts Scott and Irwin guided their lunar module to a safe touchdown on the moon four days later. Named *Falcon*, this modified lunar module was able to carry a heavier payload, including the first Lunar Roving Vehicle (LRV). The vehicle was a battery-operated moon buggy designed to transport Scott and Irwin long distances along Hadley Rille, a narrow valley near the Apennine Mountains.

During their three lunar excursions, Scott and Irwin rode seventeen miles in the LRV, giving them the freedom to conduct more extensive geological surveys. One of the most important samples collected during the mission was the Genesis Rock, believed to be a piece of lunar crust from about the time of the moon's origin.

The astronauts stayed on the lunar surface for almost sixty-seven hours. Worden, who piloted *Endeavour* in lunar orbit while Scott and Irwin were on the moon, conducted the first Apollo spacewalk on the way back to Earth in order to retrieve film from the side of *Endeavour*'s service module. The crew landed in the Pacific Ocean on August 7, 1971, to successfully complete the mission.

The astronauts carried five 70-mm cameras into space with nineteen magazines of film. They captured 1,518 black-and-white images 1,018 in color and 105 on ultraviolet film.

PREVIOUS SPREAD (LEFT)

The Apollo 15 crew aboard the NASA Motor Vessel Retriever talk with Navy Lieutenant Fred W. Schmidt, the assigned lead diver for recovery operations on May 2, 1971. The crew—from left to right, Alfred Worden, James Irwin, and David Scott—was in the Gulf of Mexico for training.

OPPOSITE

A frame from a panorama taken during the second EVA by James Irwin shows the lunar module *Falcon* at the foot of the Apennine mountain range.

■ Apollo 15—Irwin: "Dave, I'm reminded of a favorite biblical passage from Psalms. 'I look unto the hills, from whence cometh my help.' But of course, we get quite a bit from Houston, too."

—DAVID R. SCOTT

RIGHT

This frame from Irwin's panorama is centered on Mount Hadley, rising about 4,765 feet above the plain, which is in full sun. Note the well-defined imprints made by the lunar rover's tires.

OPPOSITE
The LRV is photographed alone against a desolate lunar background during the third EVA at the mission's Hadley-Apennine landing site. The west edge of Mount Hadley is at the upper right edge of the picture.

CLOCKWISE FROM RIGHT
Scott is seated in the LRV during the first EVA. Photograph by James B. Irwin.

Scott is at the back of the LRV examining the vise. He has a core with him, lying horizontally along the back of the rover. Photograph by James B. Irwin.

The LRV from the rear. Photograph by James B. Irwin.

Commander Scott at the LRV.

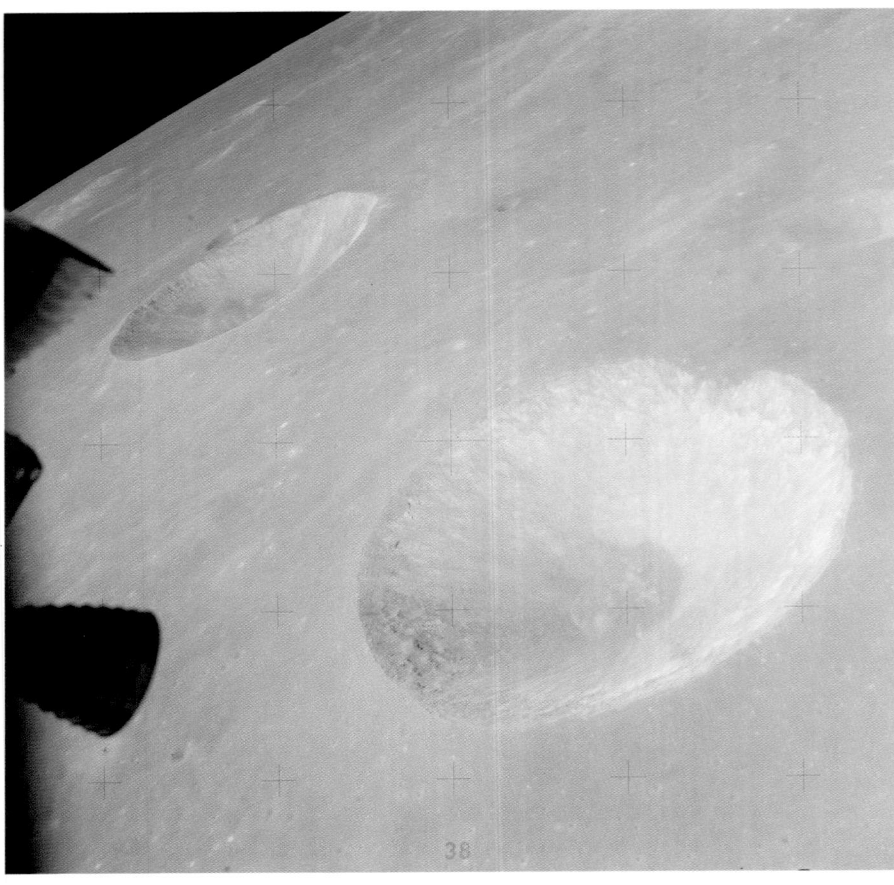

ABOVE
Craters Carmichael and Hill in a photograph taken by Dave Scott from the lunar module. An LM thruster is visible at the left.

OPPOSITE
View of the Montes Agricola taken from the command module by Alfred Worden.

■ This was one of the more spectacular photos taken during lunar orbit. I was passing over the area of the moon where the sun was setting, and the shadows were stretching and revealing every small variation in the lunar features that produced a format for analyzing the height of those objects in very precise detail. This line was, of course, the terminator, or the place where the sun was just setting. Many of the photo magazines used these photos to try to prove that some of the objects were man-made.

However, knowing the sun angle allowed the geologists to calculate the height and horizontal dimension of the objects, and their general formation, including how the ancient geologic forces created them. This particular area looks like a pressure ridge formed by extruded lava and the resulting flow of molten lava. A sure sign that the moon had, at some point in its history, volcanic activity much as we have here on Earth.

—ALFRED M. WORDEN

PRESS

APOLLO 16

JOHN W. YOUNG
THOMAS K. "KEN" MATTINGLY II
CHARLES M. DUKE JR.

Apollo 16 was the fifth mission to successfully land astronauts on the moon and the second to conduct a longer, more science-oriented lunar expedition.

The crew's powerful Saturn V rocket was launched from the Kennedy Space Center on April 16, 1972. Commander John Young, making his second trip to the moon, was joined by astronauts Charlie Duke and Thomas K. "Ken" Mattingly, who was pulled from his Apollo 13 assignment after being exposed to the measles. The crew named their command module *Casper* after the ghostly appearance of the astronauts' white spacesuits. The lunar module was dubbed *Orion*. For the second time, the lunar module carried a rover to the moon's surface.

A problem occurred after *Casper* undocked from *Orion* in lunar orbit. As the lunar module was descending to the moon's Descartes Highlands, Mattingly reported a malfunction in the system that controlled the steering of *Orion's* main engine.

Young and Duke had another problem aboard *Casper*. The lunar module had a glitch that pressurized the propellant tanks to dangerously high levels. The crew relieved the pressure by firing the thrusters, but the tactic cost them precious fuel. The two spacecraft flew in formation above the moon as NASA managers in mission control debated whether or not to scrub the lunar landing.

The decision was eventually made to proceed, and on April 21, 1972, *Orion* landed safely and on target. Of the six Apollo landings, this was the only one to explore the lunar highlands.

Young and Duke spent more than twenty hours walking and driving on the moon during three moon walks. Among the records set during this mission was a lunar land-speed record of more than eleven miles per hour in the moon buggy. The crew safely landed in the Pacific Ocean on April 27, 1972.

The Apollo 16 mission carried four 70-mm cameras and twenty-two magazines of film to the moon. The astronauts captured a total of 2,808 images: 1,224 in black-and-white, 1,501 in color, and eighty-three on ultraviolet film.

16

PREVIOUS SPREAD
The Apollo 16 homecoming ceremonies are underway on the deck of the USS *Ticonderoga* shortly after splashdown on April 27, 1972. John W. Young is standing at the microphone. Standing behind him are Charles M. Duke Jr. and Thomas K. "Ken" Mattingly II.

OPPOSITE
John W. Young leaps from the lunar surface as he salutes the U.S. flag at the Descartes landing site during the first Apollo 16 EVA, April 21, 1972. The lunar module *Orion* is on the left and the LRV is parked beside the lander. Photograph by Charles M. Duke Jr.

■ I am saluting the flag on the moon. This shows the advantage of lunar gravity. I weighed, with my suit and backpack, about 360 earth pounds, but only 60 pounds in the 1/6th gravity of the moon. The "Jumping Salute" photo is one of the most famous lunar pictures, as it shows how nice it will be to live and work on the moon (by 2020, with any luck!). Charlie Duke took this photo with his chest-mounted 250-mm Hasselblad 250 camera.

—JOHN W. YOUNG

RIGHT
Duke stands in the shadow of *Orion* behind the ultraviolet camera. This photograph was taken by John Young during the mission's second EVA on April 22, 1972.

OVERLEAF
Two frames from a panoramic view of Station 11 taken by John Young during the mission's third EVA on April 23, 1972. Duke is by the rover, which is just visible at the edge of the rim crest of North Ray Crater.

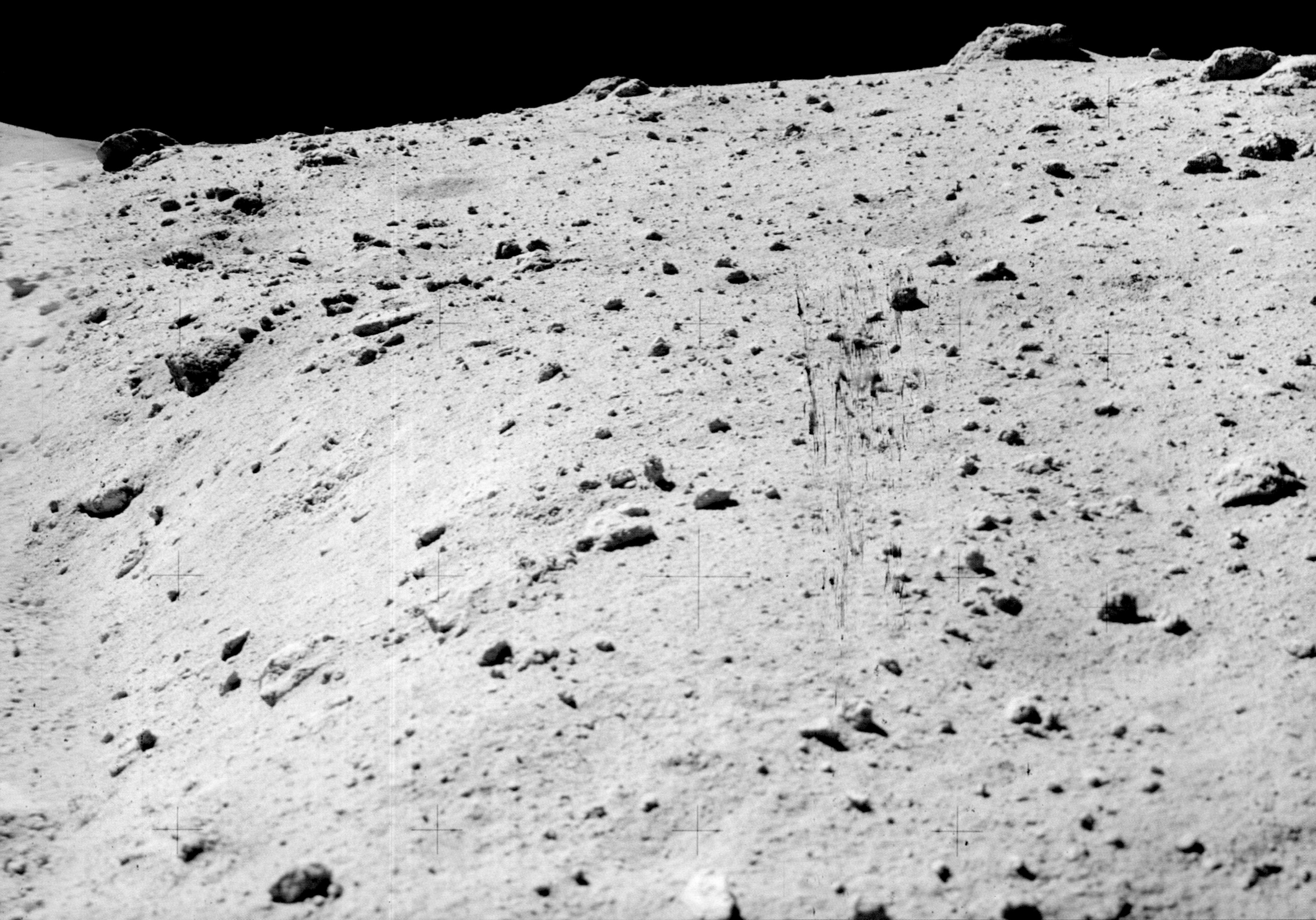

ABOVE
A photograph taken by John Young shows the front of the LRV with the dust brush and the open Lunar Communications Relay Unit blankets during the second EVA. Duke is working next to the rover.

OPPOSITE
Duke's shot of his family photograph on the surface of the moon, taken during the third EVA.

■ As I was training for my flight to the moon on Apollo 16, it was important for me to get my family involved. As a family, we went to KSC for the second roll out of our Saturn V. The family watched some training exercises and I would talk to them about what I would be doing. During one of these discussions, we thought it would be great to leave a photo of the family on the moon. One of my friends at NASA, Ludy Benjamin, came over and took the photo in our backyard. On the back of the photo I wrote, "This is the family of Astronaut Charlie Duke from planet Earth who landed on the moon on April 20, 1972." We all signed it. I got permission from NASA to carry the photo with me and to leave it on the moon. During our last EVA, I dropped the photo on the moon and took a picture of the photo. It was a very special moment.

—CHARLES M. DUKE JR.

OPPOSITE
Thomas K. Mattingly II, command module pilot, performs an EVA during the Apollo 16 trans-Earth coast, April 25, 1972, assisted by Duke in the foreground. This view is a frame from motion picture film exposed by a 16-mm Maurer camera.

■ Lunar missions can be described as a ceaseless stream of unimaginable experiences. This picture is a reminder of one of these; an emotionally overwhelming sensation of space. It was taken halfway between the Earth and moon during our return. My job was to crawl along the side of the service module and retrieve film for return to Earth in the command module. The spacecraft was oriented with the sun at my back. Having spent a week floating around, watching the Earth rise over the surface of the moon and marveling at the sight of millions of visible stars, I thought it would be good to take one last look around.

I positioned myself to look up, left, and right, from the surface of the service module. There was nothing to see except our silver spacecraft with Charlie in the hatch. No colorful Earth, no white moon, not one star; nothing but us and the sun. Without visual cues, even the familiar voice from mission control did not break the profound sense of detachment and appreciation for the true meaning of the expression "deep space". Yet, once I raised the sun visor, the stars and a sense of normalcy returned.

—THOMAS K. "KEN" MATTINGLY II

11
MIRROR MOUNT

APOLLO 17

EUGENE A. CERNAN
RONALD E. EVANS
HARRISON H. "JACK" SCHMITT

Apollo 17 would be the program's final journey to the moon. The mission was a spectacular conclusion to the first era of human space exploration. One of the last two men to walk on the moon was a first for NASA—a true scientist, geologist Jack Schmitt.

Apollo 17 was the only night launch of the program, lifting off at 12:33 AM, December 7, 1972. The mission's destination, the Taurus-Littrow region of the moon near the rim of the Serenitatis Basin, was a geologist's dream. It had steep-walled valleys with large boulders at their base. The area gave NASA an opportunity to sample both young volcanic rock and older mountainous wall material at the same location.

Joining Schmitt on the lunar surface was veteran Apollo astronaut Gene Cernan. It was his second trip to the moon. During Apollo 10, Cernan flew the lunar module to within fifty thousand feet of the surface. This time, on December 11, 1972, Cernan took his second moon lander, named *Challenger*, all the way down. In lunar orbit above Cernan and Schmitt was astronaut Ron Evans, pilot of the command module *America*.

Like the two preceding Apollo missions, there were three moon walks. The astronauts used the lunar rover to travel almost nineteen miles, collecting 243 pounds of rock and soil. Apollo 17 suffered the first extraterrestrial fender bender when Cernan accidentally ripped one of the rover's bumpers with a hammer. He used a plastic map to make repairs. Cernan and Schmitt spent twenty-two hours walking and driving on the moon, out of almost seventy-five hours on the lunar surface.

Challenger lifted off from the moon's surface December 14, 1972. On the lunar lander's decent stage, the astronauts left behind a plaque that reads, "Here Man completed his first exploration of the Moon, December 1972 A.D. May the spirit of peace in which we came be reflected in the lives of all mankind."

The Apollo program's lunar missions ended December 19, 1972, when the crew splashed down safely in the Pacific Ocean. They had carried four 70-mm cameras and twenty-three magazines of film on this final mission to the moon, capturing 3,584 images; 1,645 in black-and-white and 1,939 in color.

OPPOSITE
Eugene Cernan is photographed inside the lunar module *Challenger* on the moon following the second EVA on December 12, 1972. Note lunar dust on his suit. Photograph by Harrison H. "Jack" Schmitt.

TOP
Jack Schmitt still has his Apollo growth of beard in this portrait taken aboard the command module *America* on the return to Earth.

BOTTOM
Ronald Evans, right, with a clean-shaven Schmitt, is seen in a photo taken by Gene Cernan on the return to Earth.

ABOVE
Gene Cernan is holding the lower corner of the American flag during the mission's first EVA, December 12, 1972. Photograph by Harrison H. "Jack" Schmitt.

■ This photo, I believe, captures it all—mankind, the moon, the Earth, the blackness and endlessness of time and space, and, perhaps of greatest importance, the flag of our nation. The legacy of Apollo is not the technology you now hold in the palm of your hand, but rather the dedication and commitment of those millions of Americans who, in troubled times, made it all possible.

—EUGENE A. CERNAN

RIGHT
Tsiolkovsky Basin photographed by Jack Schmitt from the command module *America*, December 15–16, 1972.

■ As debate over the landing site for the final Apollo mission to the moon took place in 1971, I asked a group of friends, mostly young Manned Spacecraft Center engineers, to help me consider the feasibility of landing on the lunar far side, specifically in the spectacular, large basin named Tsiolkovsky. Previously, after the Apollo 13 rescue, I had recommended informally that the four Apollo missions after Apollo 14 land at publicly exciting places, sites to which most lunar scientists had always wanted to go, anyway.

These locations consisted of Tycho Crater in the South, Orientale Basin in the West, a site near permanent shadow in the North, and Tsiolkovsky on the far side as the culmination of humankind's first great adventure in space exploration—Apollo. Needless to say, little interest appeared in this proposal as NASA worked to recover from Apollo 13's exploding oxygen tank and the nerve-racking rescue of the crew. As Apollo 17 planning began, however, I thought Tsiolkovsky was worth one more try. We had good photographic coverage for mission planning and, with landmark tracking to fix our landing point relative to the spacecraft orbit, a far-side landing could be done, provided a far-side-to-Earth communication link could be established. Such a link proved to be possible if a relay satellite were placed at a Lagrangian libration point 56,000 km behind the moon. We located two relay satellites in storage, one for redundancy, and identified an appropriate launch vehicle.

Apollo 17, of course, did not land near the central peak of Tsiolkovsky. As word of our efforts leaked out, the director of the center, Christopher Kraft, stopped me in the hall one day and said, "Jack, forget this idea." I did, however, take numerous photographs of the basin, including this one from directly overhead.

—HARRISON H. "JACK" SCHMITT

OPPOSITE
Wide-angle view of the Apollo 17 Taurus-Littrow lunar landing site taken by Gene Cernan during the first EVA. To the left, in the background, is the lunar module *Challenger*. To the right in the background, is the LRV. Schmitt is visible between the two points.

ABOVE
This photograph was taken by Gene Cernan during the deployment of the Apollo Lunar Surface Experiments (ALSEP) during the first EVA. An ALSEP was carried out by each of the five missions to land on the moon after Apollo 11.

OVERLEAF
Jack Schmitt, a geologist, is photographed standing next to a huge, split boulder during the third EVA at the Taurus-Littrow landing site, December 14, 1972, in two frames from a panorama taken by Gene Cernan. The LRV is seen to the right.

PREVIOUS SPREAD (LEFT)
The crescent Earth rises above the lunar horizon in a photograph taken from the Apollo 17 command module in lunar orbit.

PREVIOUS SPREAD (RIGHT)
In this view, taken from the lunar module *Challenger*, the command module *America*, piloted by Evans, is seen preparing to rendezvous with the lander. Note the reflection of the lunar surface on the command module.

ABOVE
A view of the lunar module *Challenger* from the command module *America* during transposition and docking maneuvers on the way to the moon, December 7, 1972. The white dots surrounding *Challenger* are debris from the Saturn S-IVB stage separation.

OPPOSITE
Almost the entire coastline of Africa is clearly visible in this photograph of Earth taken by Jack Schmitt as the Apollo 17 traveled toward the moon, December 7, 1972.

■ Upon assignment to the Apollo 17 mission, I began to plan observations of the Earth while en route to the moon. The entire planet could be watched as it rotated in space during the three-and-one-half-day flight. As we left for launch pad 39 on December 6, 1972, I stuffed my space suit pocket with the latest full-Earth satellite photographs Air Force meteorologists had provided.

Once we were out of Earth orbit and on our way to the moon at 25,000 miles per hour, these low-resolution, black-and-white satellite photographs helped begin my observations of changes in weather patterns. At first, we were too close to see the full Earth; but rapidly Madagascar, then South Africa, then all of Africa dominated the view out the windows of the Command Module *America*. The views expanded quickly until the nearly full Earth filled the large hatch window and finally, not even that. This now-famous picture of Earth, taken from about 34,000 miles away, shows all of Africa, the continent of human origins and later migrations. One of a large sequence of pictures taken over three days, this image compliments page after page of weather descriptions. In spite of the personal motivations for the descriptions and photographs, when I took this picture, I could not help but be struck by the remarkable fact that humans from that now-receding blue, green, red-yellow, and white globe could take such a picture. A new migration to places elsewhere in the solar system indeed had begun. As a geologist, I also reflected on how much our home planet had endured over four and a half billion years of time, demonstrating a truly remarkable resilience to apparent adversity.

—HARRISON H. "JACK" SCHMITT

NASA

This book is dedicated to the first Apollo crew and all who gave their lives in the pursuit of space exploration.

VIRGIL I. "GUS" GRISSOM
EDWARD H. WHITE II
ROGER B. CHAFFEE

Our God-given curiosity will force us to go there ourselves because in the final analysis, only man can fully evaluate the moon in terms understandable to other men.

—GUS GRISSOM

OPPOSITE
The crew of Apollo 1 portrayed during training for their mission, January 17, 1967. Ten days later, they would die in a fire during a test. From left to right, they are Virgil I. "Gus" Grissom, Edward H. White II, and Roger B. Chaffee.

Acknowledgments

We deeply appreciate the cooperation and participation of the Apollo astronauts and the nearly four hundred thousand people who made possible the greatest achievement in the history of human exploration.

Also, we would like to acknowledge the NASA leaders who gave us the time and opportunity to work on this project, including Chief of Strategic Communications Christopher M. Shank and Assistant Administrator for Public Affairs David R. Mould.

We especially want to thank Professor Stephen Hawking and his daughter Lucy for their valuable contributions to this publication and their insights into the value of space exploration. Once again, we thank Eric Himmel and his colleagues at Abrams, among them Michelle Ishay-Cohen, Kwasi Osei, Eric Klopfer, and Alison Gervais, for turning our vision into reality.

We appreciate the help of Dick Underwood who trained the astronauts to photograph in space. Mike Gentry and Gwen Pitman in the photo departments of the Johnson Space Center and NASA Headquarters are also to be thanked for their contributions.

Finally, we would like to give special thanks to the dedicated men and women of the National Aeronautics and Space Administration for their ongoing commitment and perseverance in achieving what many consider to be unachievable. The American people can be proud of them and proud of their space program as NASA continues its ongoing mission to pioneer a new and exciting future.

—THE EDITORS

Editor: Eric Himmel
Editorial Assistant: Eric Klopfer
Art Director: Michelle Ishay
Designer: Kwasi Osei
Production Manager: Alison Gervais

Cataloging-in-Publication Data has been applied for and may be obtained from the Library of Congress. ISBN: 978-0-8109-2146-7

Printed and bound in China
10 9 8 7 6 5 4 3 2 1

harry n. abrams, inc.
a subsidiary of La Martinière Groupe
115 West 18th Street
New York, NY 10011
www.hnabooks.com

OPPOSITE
President Kennedy speaks before a crowd of 35,000 people at Rice University's football field. The following are excerpts from his speech. ". . . We set sail on this new sea because there is a new knowledge to be gained, and new rights to be won, and they must be won and used for the progress of all people. . . . Whether it will become a force for good or ill depends on man, and only if the United States occupies a position of preeminence can we help decide whether this new ocean will be a sea of peace or a new terrifying theater of war. But I do say space can be explored and mastered without feeding the fires of war, without repeating the mistakes that man has made with extending his writ around this globe of ours. . . . There is no strife, no prejudice, no national conflict in outer space as yet. Its conquest deserves the best of all mankind, and its opportunity for peaceful cooperation may never come again. But why, some say, the moon? Why choose this as our goal? And they may well ask, why climb the highest mountian? Why —thirty-five years ago—why fly the Atlantic? Why does Rice play Texas? We choose to go to the moon, we choose to go to the moon in this decade and do the other things, not because they are easy, but because they are hard, because that goal will serve to organize and measure the best of our energies and skills, because that challenge is one that we are willing to accept, one we are unwilling to postpone, and one in which we intend to win, and the others too."

OVERLEAF
A member of the Apollo 11 crew took this photograph of the moon on the journey homeward, July 21, 1969.

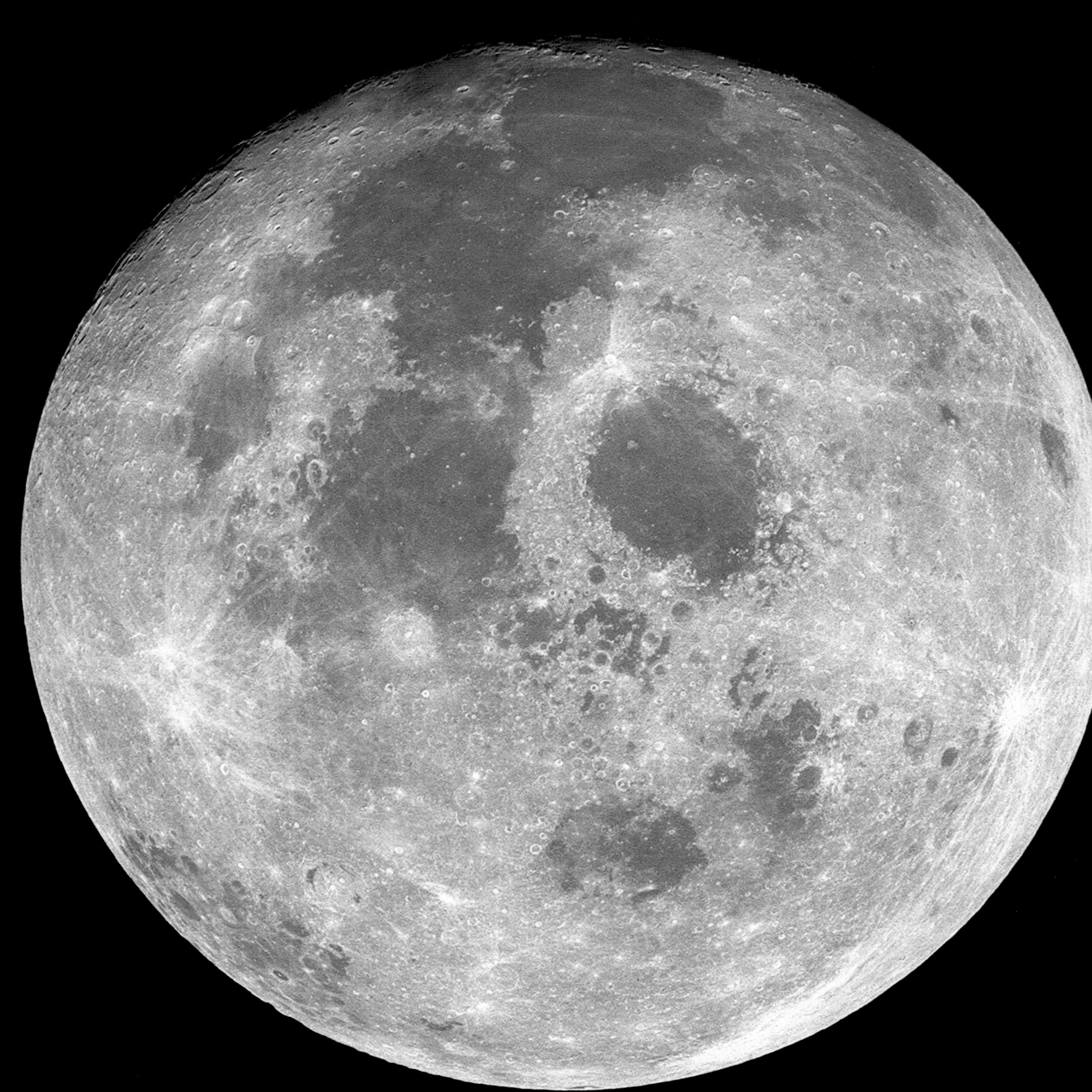